Connected and Automated Vehicles 2018

Transportation Research Record: Journal of the Transportation Research Board

Volume 2672, Issue 22

ISBN: 978-1-5108-7827-3

Printed from e-media with permission by:

Curran Associates, Inc.
57 Morehouse Lane
Red Hook, NY 12571

Some format issues inherent in the e-media version may also appear in this print version.

Published by Sage Publications USA

ISBN (Print) 978-1-5108-7827-3
ISBN 2018 Printed Set (All Issues) 978-1-5108-7735-1

Printed by Curran Associates, Inc. (2019)

Additional copies of this publication are available from:

Curran Associates, Inc.
57 Morehouse Lane
Red Hook, NY 12571 USA
Phone: 845-758-0400
Fax: 845-758-2633
Email: curran@proceedings.com
Web: www.proceedings.com

Volume 2672 Number 22 December 2018

Contents

Connected and Automated Vehicles 2018

Articles

TRANSPORTATION RESEARCH BOARD

CORE SPONSORS

Transportation Departments of the 50 States and the District of Columbia

Federal Government

U.S. Department of Transportation
- Federal Aviation Administration (FAA)
- Federal Highway Administration
- Federal Motor Carrier Safety Administration (FMCSA)
- Federal Railroad Administration (FRA)
- Federal Transit Administration
- National Highway Traffic Safety Administration (NHTSA)
- Office of the Assistant Secretary of Transportation for Research and Technology (OST-R)

U.S. Air Force Civil Engineer Center (AFCEC)
U.S. Coast Guard (USCG)
U.S. Department of Energy
U.S. Department of the Interior (DOI)
U.S. Environmental Protection Agency (EPA)

Nongovernmental Organizations

American Public Transportation Association (APTA)
Association of American Railroads (AAR)
California Air Resources Board

The National Academies of
SCIENCES • ENGINEERING • MEDICINE

The **National Academy of Sciences** was established in 1863 by an Act of Congress, signed by President Lincoln, as a private, nongovernmental institution to advise the nation on issues related to science and technology. Members are elected by their peers for outstanding contributions to research. Dr. Marcia McNutt is president.

The **National Academy of Engineering** was established in 1964 under the charter of the National Academy of Sciences to bring the practices of engineering to advising the nation. Members are elected by their peers for extraordinary contributions to engineering. Dr. C. D. Mote, Jr., is president.

The **National Academy of Medicine** (formerly the Institute of Medicine) was established in 1970 under the charter of the National Academy of Sciences to advise the nation on medical and health issues. Members are elected by their peers for distinguished contributions to medicine and health. Dr. Victor J. Dzau is president.

The three Academies work together as the National Academies of Sciences, Engineering, and Medicine to provide independent, objective analysis and advice to the nation and conduct other activities to solve complex problems and inform public policy decisions. The Academies also encourage education and research, recognize outstanding contributions to knowledge, and increase public understanding in matters of science, engineering, and medicine.

Learn more about the National Academies of Sciences, Engineering, and Medicine at **www.national-academies.org.**

The **Transportation Research Board** is one of seven major programs of the National Academies of Sciences, Engineering, and Medicine. The mission of the Transportation Research Board is to increase the benefits that transportation contributes to society by providing leadership in transportation innovation and progress through research and information exchange, conducted within a setting that is objective, interdisciplinary, and multimodal. The Board's varied activities annually engage about 7,000 engineers, scientists, and other transportation researchers and practitioners from the public and private sectors and academia, all of whom contribute their expertise in the public interest. The program is supported by state transportation departments, federal agencies including the component administrations of the U.S. Department of Transportation, and other organizations and individuals interested in the development of transportation.

Learn more about the Transportation Research Board at **www.TRB.org**.

Article

TRR
JOURNAL OF THE TRANSPORTATION RESEARCH BOARD

Transportation Research Record
2018, Vol. 2672(22) 1–9

Article reuse guidelines:
sagepub.com/journals-permissions
DOI: 10.1177/0361198118775871
journals.sagepub.com/home/trr
SAGE

Necessity of Automated Vehicle Control Customization: Experimental Results during Lane Changing

Naohisa Hashimoto[1], Simon Thompson[1], Shin Kato[1], Ali Boyali[1], and Sadayuki Tsugawa[1]

Abstract

This study investigated the necessity of automated vehicle control customization for individual drivers via a lane-changing experiment involving 35 subjects and an automated minivan. The experiment consisted of two automated driving conditions: one in which the subject was unable to override vehicle controls, the other with the option to override when the subject felt it was necessary. The automated vehicle drove at a speed of 40 km/h along three kinds of planned paths for lane changing, generated by Bezier curves; the distance required for lane changing was varied to obtain the preferred path of each subject. Various data obtained during driving, including vehicle trajectories and steering angles produced by subjects were logged. After automated driving, a questionnaire was administered to each subject. The experimental data showed that there was a statistically significant difference between comfort when the vehicle drove along the subject's preferred path, and when it drove along other paths. The results of the questionnaire indicated that 46% of the subjects preferred the planned path that most closely resembled their own. In addition, quantitative analysis of driving data found that approximately 69% of the subjects preferred an automated driving control that resembled their own. However, it was also observed that certain subjects were open to multiple types of automated vehicle control. The experimental results indicate that drivers will not necessarily accept a single type of automated vehicle control, therefore customization will be necessary to improve acceptance of automated driving.

An essential requirement for the successful introduction of automated driving to the market and society is the acceptance by drivers and passengers of automated vehicles; acceptance could potentially bring benefits such as safety, efficiency, convenience, and comfort (*1–6*). The human factors related to the control of automated vehicles will be one of the relevant aspects in promoting acceptance, but little work has been undertaken in this regard to date (*7–11*). This paper deals with human factors related to the control of automated vehicles, and presents evidence, via an experiment with subjects driving an automated minivan during lane changing, for the customization of automated vehicle controls.

Much work has been carried out in relation to driver behavior models of lane-change timing for automated vehicles using real experimental data on the highway. Do et al. investigated lane change behavior, and evaluated the validity of the proposed models in a simulation (*12*). Nilsson et al. compared manual and automated driving by employing track drivers in a simulation (*13*). Wanga et al. proposed and evaluated a decision-making model for automated vehicles that facilitated optimal lane-change times and acceleration in certain scenarios (*14*). Kinnear et al. researched lane change behavior with a focus on timing, and summarized the requirements for the assistance system (*15*). Mazer et al. analyzed the differences between older and younger drivers in several situations, including lane changes, involving more than 100 subjects (*16*). Wiedemann et al. analyzed driver's interaction on different automation levels (*17*). Griesche et al. studied the driver's preferred driving style for the design of automated vehicle functions (*18*).

In this study, Bezier curves were employed to generate the paths. There are several research studies on the use of Bezier curves. Nagasaka and Harada employed a Bezier

[1]National Institute of Advanced Industrial Science and Technology (AIST), Tokyo, Japan

Corresponding Author:
Address correspondence to Naohisa Hashimoto:
naohisa-hashimoto@aist.go.jp

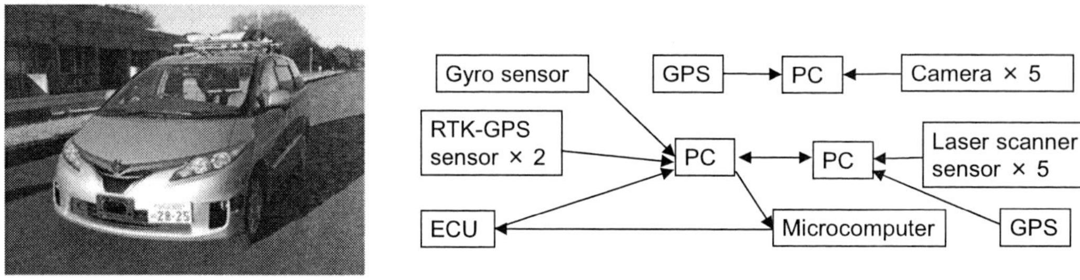

Figure 1. Photograph and system configuration of experimental vehicle.

curve for automated vehicles to generate a comfortable path (*19*). Gonzalez used this curve in considering the speed profile needed to achieve passenger comfort (*20*). Liu used a Bezier curve as a method to estimate dangerous situations in lane-change maneuvers (*21*). The rationale for a Bezier curve in this study was because several types of curves can be expressed easily by selecting a few control points; and the paths of several subjects could be categorized using the curves.

The experimental vehicle, scenario, method of categorization, results, and discussion will be presented in the following sections.

Experiment

Experimental Vehicle

Figure 1 shows the experimental vehicle and its system configuration. The vehicle was a modified commercial hybrid vehicle. It was controlled using a PC through a control area network; the system was able to obtain vehicle status information through the OBD-II port. In addition, two RTK-GPS sensors (Novatel OEM628), a GPS sensor, two PCs, a gyro sensor (Tamagawaseiki TA7265), five camera sensors for monitoring the driver, and five laser scanner sensors (ibeo LUX) for recognition of the surroundings were installed. In this study, mainly the two RTK-GPS sensors, the gyro sensor, and vehicle status information were used for realizing the automated driving functions. Data were processed using the Extended Kalman Filter. A target path for automated vehicle control was generated using localization data from the RTK-GPS (*22*).

Subjects for the Experiment

Thirty-five subjects (18 men and 17 women) were employed for the experiments, and were instructed about the objectives before driving. Subjects were asked questions about personal characteristics and driving-related topics (gender, age, history of driving, possession of driver's license, traffic accidents, driving frequency, usual driving distance, primary purpose of driving, vehicle type for everyday driving, driving habits), and were further administered the Driving Style Questionnaire and the Workload Sensitivity Questionnaire developed by Ishibashi et al. (*23*). The ages of the subjects ranged from 28 to 67 (average: 44.4 years). All subjects had a driver's license.

Generation of Paths for Automated Lane Changing

Before the description of the experiment, the generation of paths for the automated vehicle during lane changing will be introduced. Three kinds of path along which the automated vehicle drives during lane changing were generated.

Several methods were proposed to generate a path for an automated vehicle. In this study, the Bezier curve, which Pierre Bézier originally proposed for making an arbitrary curve with a small amount of data, was employed (*24*). Two parameters (the Bezier pattern and the distance of the path) were used to categorize a path for each subject. To change a Bezier curve, three sets of Bezier patterns were selected, and the distances were selected at five intervals. A fourth-order Bezier curve was employed: three control points in addition to start and goal points were necessary to use a fourth-order curve, as shown in Figure 2. In this study, three patterns (A, C, and E) were used by changing the relative positions of the three control points. The three control points are shown as three blue dots (the middle three points among the five points) in Figure 2. X(i), for deciding the relative positions of the points in Figure 2, was determined as follows:

$$X_A(i) = 1 \times D/8$$

$$X_C(i) = D/4$$

$$X_E(i) = 3 \times D/8$$

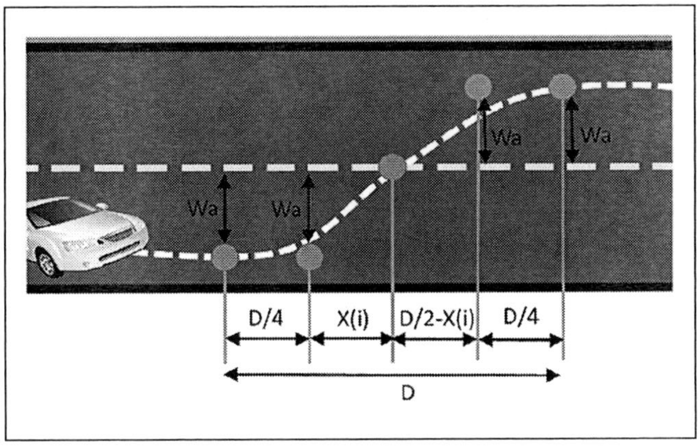

Figure 2. Path generation method using a Bezier curve. The blue dots represent control points in the Bezier curve. The green dots represent start and goal points.

The distance between the lane marker and each point (Wa) in Figure 2 is identical. In addition, the distance (D) of the path was varied from 50 m to 70 m at 5 m intervals. Distance traveled during the overtaking maneuver is one way to characterize the driving path. Before entering and after completing the overtaking path, the vehicle moves in a straight line. By changing two types of parameters and using the root mean square (RMS) method, the nearest trajectory of each subject was obtained. Figure 3*a* illustrates the three patterns of generated paths; these three patterns were used in the experiment. The feature of pattern A is that the turning radius is small at the beginning of the lane-changing maneuver and large at the end. On the contrary, pattern E's turning radius is large at the beginning and small at the end. Pattern C is the intermediate between A and E. In an experimental study with questionnaires, it would have been difficult for subjects to remember several patterns, thus only the three were employed. For example, the paths of a subject are C70 (shown in Figure 3*b*), E50 and A50.

For comparison, the subject's alternative path was generated – the one which was most different from the subject's path. The subject's alternative path could be simply generated by the following two methods.

1. Control point

If the control point of the subject's path is A, the control point of the alternative path is E. If the control point is C, the alternative path control point is E. If the control point is E, the alternative path control point is A.

2. Distance of a path (D)

If the distance of a path is under 55 m, the alternative path is set to 70 m. If the distance of a path is over 60 m, the alternative path is set to 50 m. For example, if the subject's path is A55, the subject's alternative path is E70. If the subject's path is C70, the subject's alternative path is E50. Using the proposed method, the subject's alternative path could be extremely different from the subject's original path.

Experimental Site

The experiment was conducted on the test track at the National Institute of Advanced Industrial Science and Technology (AIST) in Tsukuba, Japan. To prevent unintended effects, no other vehicles were allowed on the test track during the experiment. All driving was conducted during the day in sunny conditions for consistency. The experimental course was a three-lane road. An object was placed in the left lane. The vehicle traveled in the

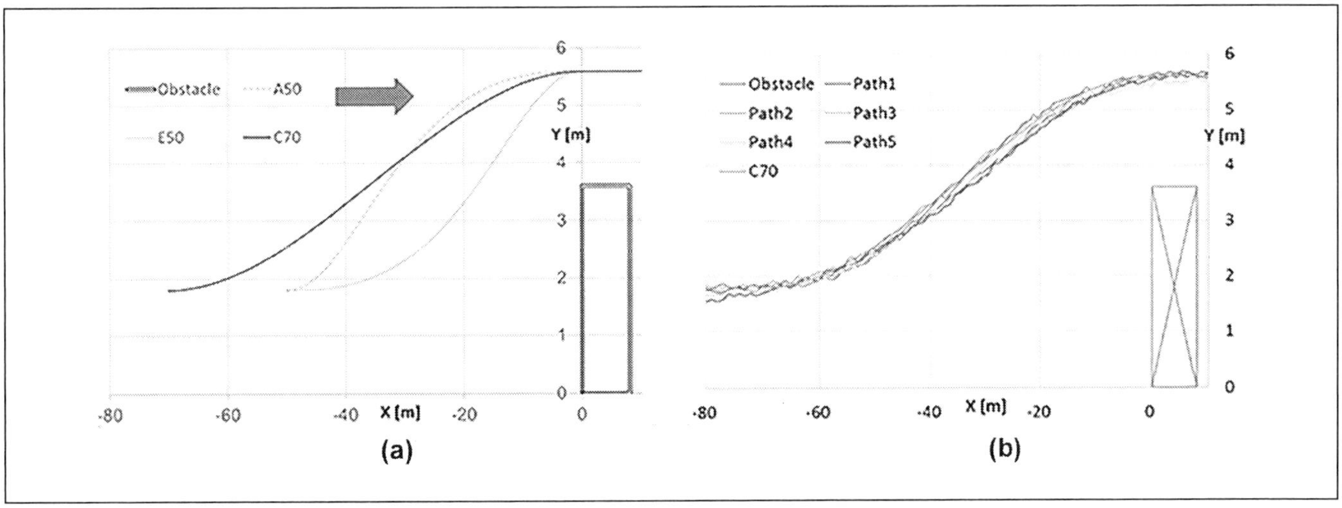

Figure 3. Generated paths: (*a*) three patterns of generated paths (the vehicle travels in the direction of the arrow) and (*b*) path example of one subject (this path was categorized as C70).

direction of the arrow, and the object was regarded as an obstacle (see Figure 3).

Scenario of Experiment

During the driving experiments each subject sat in the driver's seat in the vehicle, and, for safety reasons, a staff member sat in the passenger seat. Subjects were asked to stay in the left-hand lane and change lanes normally to avoid an obstacle. One static obstacle was set on the course. The velocity of the vehicle was maintained at 11.1 m/s (40 km/h).

Step 1: Manual Driving by Subjects. After the completion of the questionnaires, the Step 1 manual driving experiment was conducted. In this experiment, driving data for lane change maneuvers were obtained to categorize each subject's driving style during lane changes. Each subject drove 10 times for this experiment. The data for the first, second, and last driving runs were not used because the data for these driving runs were largely different from the employed data; data for five runs out of the remaining seven were selected using RMS. It was presumed that the experimental vehicle was different from the vehicle subjects usually drove, and they needed to familiarize themselves with the vehicle.

Step 2: Automated Driving without Overriding by Subjects. Subsequently, the Step 2 automated-driving experiment without overriding was conducted. During the experiment, each subject was asked to sit in the driver's seat and not override the controls of the experimental vehicle. The automated vehicle drove with four path patterns of lane changes: C70, A50, E50, and the subject's path that was acquired during Step 1. In the experiment, the order of the four path patterns was randomly set. Thus, the subject was unable to recognize which path corresponded to the four path patterns. The automated vehicle traveled along the target path autonomously. Three lane-change maneuvers were performed in each path. Thus, the subjects experienced each path three times. For safety, a staff member sat in the passenger seat, and every subject had completed a training course to prepare for possible emergencies prior to participation. In addition, each subject was given the responsibility of controlling the experimental vehicle during potential emergencies, such as avoiding an accident. However, in the experiments, no emergency occurred. According to the Society of Automotive Engineers' (SAE) definition of automated driving levels, the level of the current experiments would be SAE Level 3 (*25*).

The automated driving without overriding experiment was followed by completion of questions about the driver's most comfortable path, their most uncomfortable path, and their reasoning.

Step 3: Automated Driving with Overriding by Subjects. Finally, the Step 3 experiment, automated driving with overriding, was conducted. In the experiment, each subject was asked to sit in the driver's seat and had the opportunity to override the experimental vehicle if the trajectory path was uncomfortable. Drivers were able to override the steering wheel because the force of the steering actuator was less than that of the drivers in this experimental vehicle, and the system did not stop controlling even if the drivers override. Five path patterns of lane changes were used for each subject. In the experiment, the order of the five path patterns was randomly set. Thus, the subject was unable to recognize which path corresponded to the five path patterns. The automated vehicle drove with five path patterns of lane changes: C70, A50, E50, the subject's path that was acquired during Step 1, and the subject's alternative path. The automated vehicle traveled along the target path autonomously. Three lane-change maneuvers were performed in each path. Thus, the subjects experienced each path three times. Safety conditions were the same as for the experiment without overriding.

Analysis of the Data Obtained by the Experiment

Analysis of the Subject Responses

The Most Comfortable Path. Figure 4 shows the responses to the question "Which was the most comfortable path on lane changing?" This figure illustrates that, due to large variations, it was not possible to group the subjects' paths into a single category.

Figure 5 shows the results of the experiment without overriding (Step 2) in relation to the most comfortable path. The most comfortable paths for the subjects include not only their own paths but also other artificially generated paths. The figure also shows gender preferences for the paths; male subjects tend to prefer their own paths.

Figure 6*a* shows the results for the question "Why did you choose the most comfortable path?" The three answers (similar to own path, high sense of safety, and smooth path or steering) in Figure 6*a* are not independent of each other because this question was not a selective question; this result indicates that 61% of the subjects preferred a path similar to their own, and it is presumed that they preferred their own paths when riding in an automated vehicle. In addition, this result indicates that it could be difficult to improve levels of acceptance of automated driving using a single common algorithm of automated vehicle control. Furthermore,

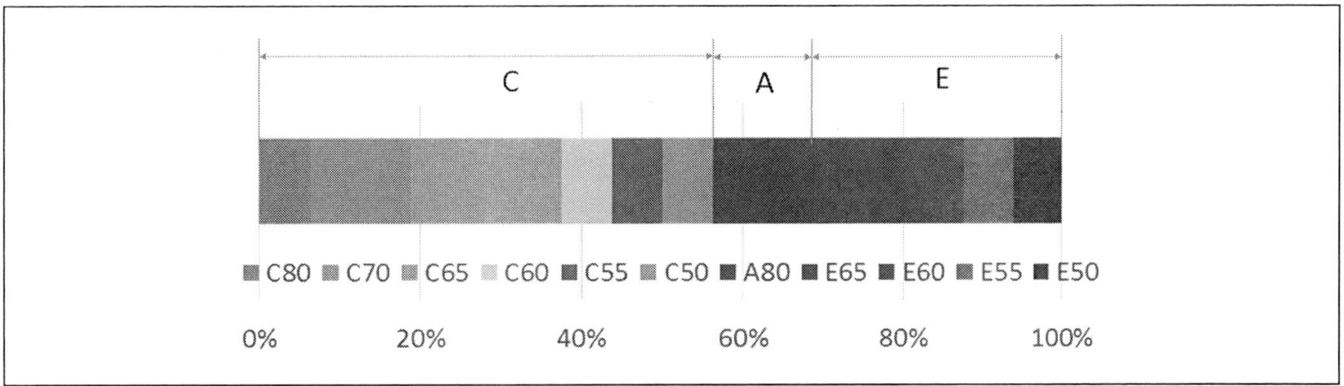

Figure 4. Questionnaire result: breakdown of the subject's paths for the question "Which was the most comfortable path on lane change?"

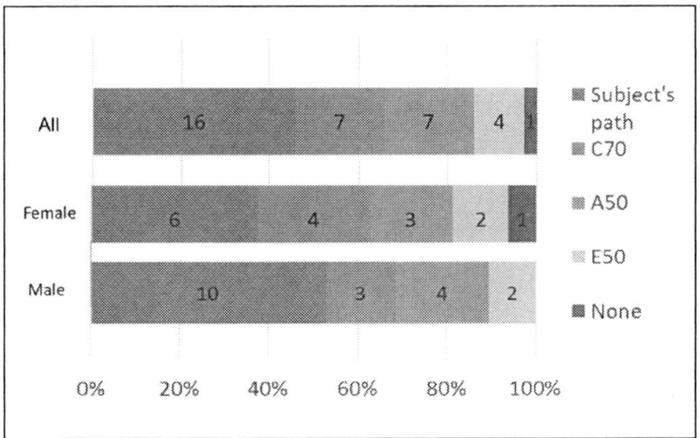

Figure 5. Questionnaire result for the question "Which was the most comfortable path on lane change?"

Bezier curves are an appropriate candidate to design a planned trajectory for automated vehicles.

The Least Comfortable Path. Figure 7 illustrates responses to the question "Which was the least comfortable path on lane changing?" Figure 7 indicates that the subjects' paths do not have the same tendency and that they have significant variations. Figure 8 illustrates the results for the least comfortable path. These results indicate that a substantially low number of subjects selected their own path. Figure 8 also indicates the answers classified by gender. These results showed the same tendency as the gender comparison results shown in Figure 5. Figure 6*b* shows the subjects' reasons in response to the question "Why did you choose the least comfortable path?" Forty-four percent of the subjects felt uncomfortable with the timing of the automated control, although all paths were sufficiently safe to maintain control of the vehicle. It is presumed that the answer "Similar to own path" in Figure 6*a* corresponds to the answer "Too late timing" and "Too early timing" in Figure 6*b*; and the answer "Smooth path or steering" corresponds to the answer "Unsmooth path

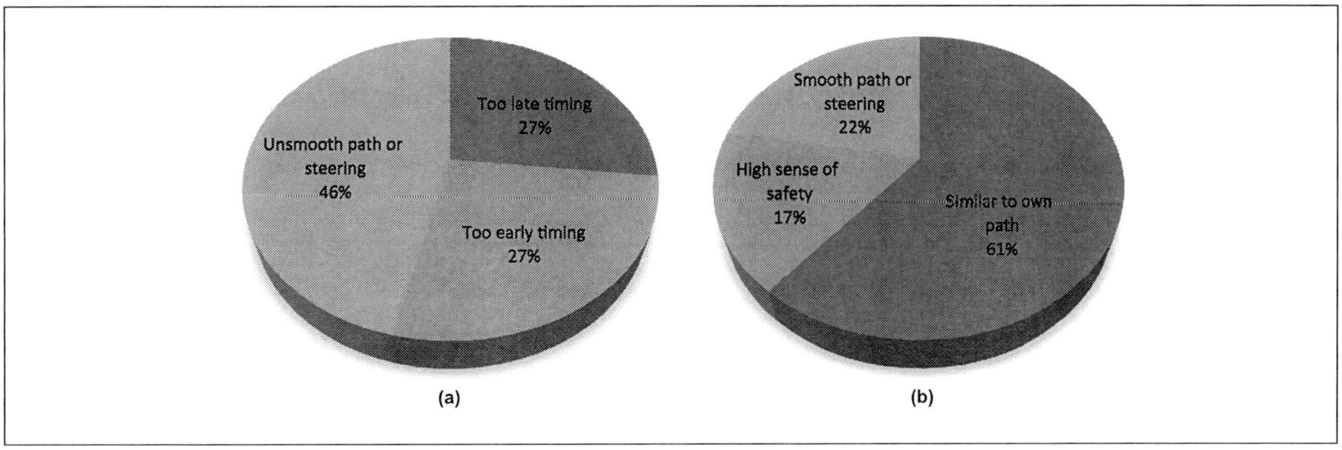

Figure 6. Reasons for path selection: (*a*) result for the question "Why did you choose the most comfortable path?"; (*b*) result for the question "Why did you choose the least comfortable path?"

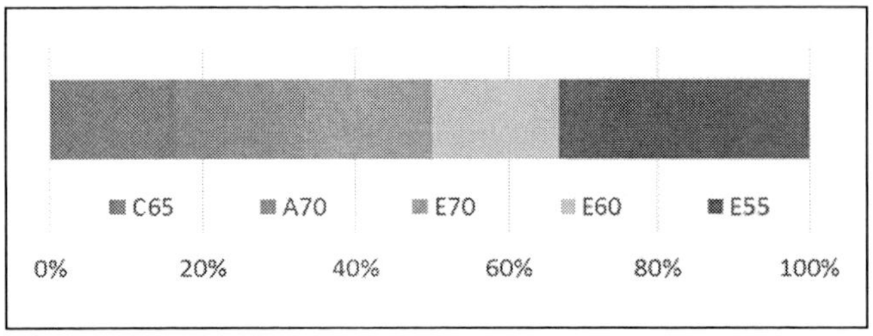

Figure 7. Questionnaire result: breakdown of the subjects' paths for the question "Which was the least comfortable path on lane change?"

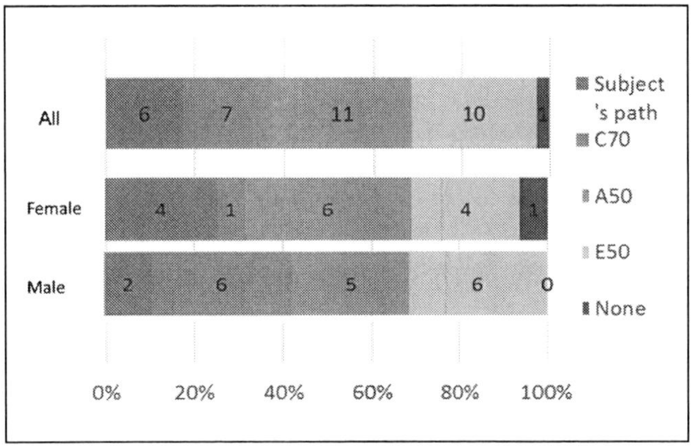

Figure 8. Questionnaire result for the question "Which was the least comfortable path on lane change?"

or steering." Thus, both the timing and the path pattern are important considerations in customizing vehicle controls according to driver style. Moreover, it is easy to decide the parameters of the timing and the path pattern using the proposed categorization method with the distance of a path and the Bezier pattern.

Quantitative Analysis of the Driving Data

With regard to the experiments involving overriding, as explained in the previous section, not only the questionnaire results but also the amount (De) of overriding was calculated as a quantitative rating, shown in Figure 9. De is calculated from the path deviation between the path with overriding and that without overriding in the Y axis in (Figure 9). RMS of De in Figure 9 was employed to evaluate the amount of overriding during lane changing; it was calculated along the path between the start point of the path and the start point of the obstacle, the area designated for lane changing. Figure 10 shows the distribution of the RMS values. From the results, it is presumed that there are differences between a subject's path and their alternative path, and differences between a subject's path and the other paths. In order to find significant differences among the paths, a *t*-test was applied (*26*). The *t*-value was calculated as follows:

$$t = \frac{\overline{X_1} - \overline{X_2}}{S\overline{X_1} - \overline{X_2}}$$

where

$$S\overline{X_1} - \overline{X_2} = \sqrt{\frac{S_1^2}{n_1} + \frac{S_2^2}{n_2}}$$

RMS values considering the amount of overriding based on the differences between the subject's path and others

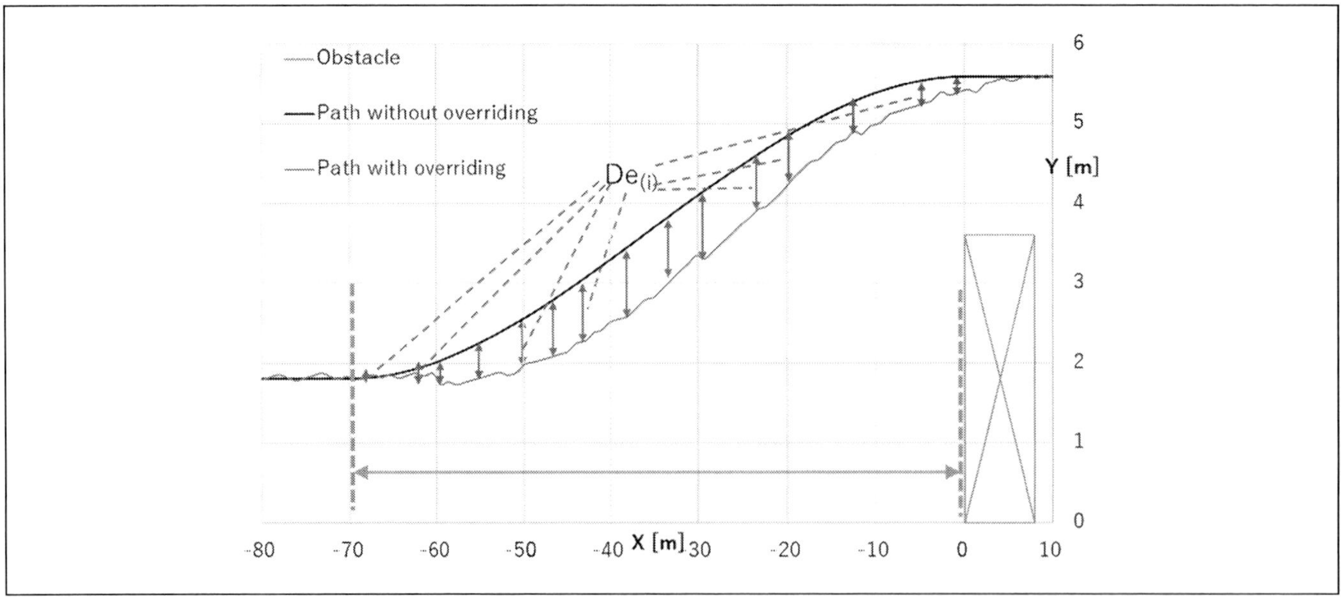

Figure 9. Amount of overriding (green line denotes calculated area).

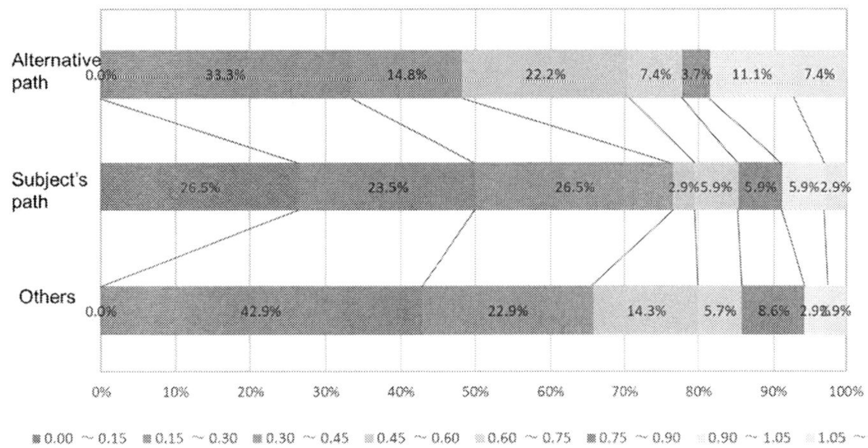

Figure 10. Experimental result: RMS values of lateral deviation of trajectories between "with overriding" and "without overriding."

Table 1. Experimental Result for the *t*-Test

	Between subject's path and the alternative path	Between subject's path and others
t-value	3.1141	2.8858
Degrees of freedom	208	489

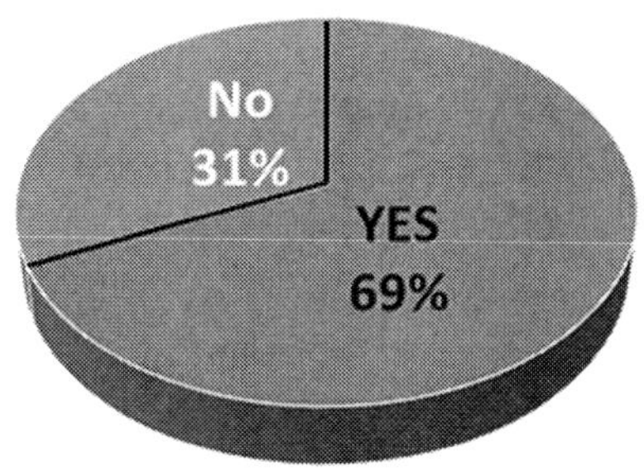

Figure 11. Questionnaire result for "Do you prefer the automated vehicle that is similar to your driving control?"

were different when the significance level was set at 1%, as shown in Table 1. Furthermore, there was an even greater difference between the RMS values of the amount of overriding based on the differences between the subject's path and the subject's alternative path when the significance level was set at 1%, as shown in Table 1. Thus, the results indicate that the subjects tended to select their own path unconsciously. Figure 11 shows the questionnaire results for the question "Do you prefer the automated vehicle that is similar to your driving control?" Approximately 69% of the subjects wanted to use the automated vehicle control that was similar to their control. This indicates that the appropriate automated vehicle control should be similar to that of an individual driver. Thus, the automated vehicle system may need to incorporate a learning function—to acquire data about driver style and preferences—before automated driving is activated, or an adapting function to match the driver's level of control by considering the amount of overriding. One of the most significant barriers to the introduction of automated driving is its acceptance, and it will be necessary for automated vehicle developers to consider both learning and adaptive controls.

Conclusions

The objective of this study was to investigate a driver adaptive path for automated vehicles. The conditions were assumed with regard to lane changing when avoiding an obstacle. Out of several possible methods for generating paths for automated driving, this study employed the Bezier curve method.

Analysis of subject responses found that 46% of the subjects considered their own path the most appropriate. The driving data indicate that drivers tend to select their own path unconsciously. In addition, approximately

70% of the subjects preferred the automated driving control that was similar to their own control.

It will be difficult to increase acceptance of automated driving using a single common automated-vehicle algorithm. Further, the control algorithms of automated vehicles must be adaptive and tunable to each driver.

Acknowledgments

This work was supported by the NEDO AI Project (P15009).

Author Contributions

The authors confirm contributions to the paper as follows: study conception and design: N.H., S.K., S.T.; data collection: N.H., S.T.; analysis and interpretation of results: N.H., A.B., S.K., S.T.; draft manuscript preparation: N.H., A.B. All authors reviewed the results and approved the final version of the manuscript.

References

1. United States Department of Transportation. *Automated Vehicle Research*. Office of the Assistant Secretary for Research and Technology (OST-R), Washington, D.C.http://www.its.dot.gov/automated_vehicle/index.htm. Accessed April, 2017.
2. ITS Japan. *2nd SIP-adus Workshop on Connected and Automated Driving Systems 2015*. Tokyo International Exchange Center, Tokyo, Japan. http://www.its-jp.org/english/files/2015/04/SIP_Worlshop2015_leaflet-s_e_20150326.pdf. Accessed April, 2017.
3. European Commission. *IOT Large Scale Pilot 5 Autonomous Vehicles in a Connected Environment*, 2015.
4. Aeberhard, M., S. Rauch, M. Bahram, G. Tanzmeister, J. Thomas, Y. Pilat, F. Homm, W. Huber, and N. Kaempchen. Experience, Results and Lessons Learned from Automated Driving on Germany's Highways. *IEEE Intelligent Transportation Systems Magazine*, Vol. 7, No. 1, 2015, pp. 42–57.
5. Shladover, S. Challenges to Evaluation of CO_2 Impacts of Intelligent Transportation Systems. *Proceedings of 2011 IEEE Integrated and Sustainable Transportation System*, 2011, pp.189–194.
6. Tsugawa, S., S. Jeschke, and S. Shladover. A Review of Truck Platooning Projects for Energy Savings. *IEEE Transaction of Intelligent Vehicles*, Vol. 1, No. 1, 2016, pp. 68–77.
7. Hoeger, R., A. Amditis, M. Kunert, A. Hoess, F. Flemisch, H. Krueger, A. Bartels, A. Beutner, and K. Pagle. Highly Automated Vehicles for Intelligent Transport: HAVEit Approach. *Proc., 15th ITS World Congress*, New York City, NY, 2008.
8. Flemischab, F., K. Benglerc, H. Bubbc, H. Winnerd, and R. Brudere. Towards Cooperative Guidance and Control of Highly Automated Vehicles: H-Mode and Conduct-by-Wire. *Journal of Ergonomics*, Vol. 57, No. 3, 2014, pp. 353–360.
9. Boer, E., and P. Spyridakos. Control-Theoretic Attention-Switching Driver Model. *Proc., International Conference on Traffic and Transport Psychology*, Brisbane, Australia, 2016.
10. Omae, M., N. Hashimoto, T. Sugamoto, and H. Shimizu. Measurement of Driver's Reaction Time to Failure of Steering Controller During Automatic Driving. *Review of Automotive Engineering*, Vol. 26, No. 2, 2005, pp. 213–215.
11. Fitch, G., R. Kiefer, J. Hankey, and B. Kleiner. Toward Developing an Approach for Alerting Drivers to the Direction of a Crash Threat. *The Journal of the Human Factors and Ergonomics Society*, Vol. 49, No. 4, 2007, pp. 710–720.
12. Do, Q., H. Tehrani, S. Mita, M. Egawa, K. Muto, and K. Yoneda. Human Drivers Based Active-Passive Model for Automated Lane Change. *IEEE Intelligent Transportation Systems Magazine*, Vol. 9, No. 1, 2017, pp. 42–56.
13. Nilsson, P., L. Laine, and B. Jacobson. A Simulator Study Comparing Characteristics of Manual And Automated Driving During Lane Changes of Long Combination Vehicles. *IEEE Transactions on Intelligent Transportation Systems*, Vol. PP, No. 99, 2017, pp. 1–11.
14. Wanga, M., W. Daamena, and R. Happeea. Optimal Lane Change Times and Accelerations of Autonomous and Connected Vehicles. Presented at 95th Annual Meeting of the Transportation Research Board, Washington, D.C., 2016.
15. Kinnear, N., S. Helman, J. Weekley, and A. Stevens. A Consideration of Driver Inattention and Technology. *Proc., International Conference on Traffic and Transport Psychology*, Brisbane, Australia, 2016.
16. Mazer, B., I. Gelinas, B. Vrkljan, T. Chen, K. Eliasz, S. Koppel, L. Charlton, M. Porter, and S. Marshall. Comparison of Naturalistic Driving Behaviors of Older and Low-risk Younger Drivers. *Proc., International Conference on Traffic and Transport Psychology*, Brisbane, Australia, 2016.
17. Wiedemann, K., N. Schömig, C. Mai, F. Naujoks, and A. Neukum. Driver's Monitoring Behavior and Interaction with Non-driving Related Tasks During Driving on Different Automation Levels. *Proc., 6th International Conference on Applied Human Factors and Ergonomics*, Las Vegas, NV, 2015.
18. Griesche, S., E. Nicolay, D. Assmann, M. Dotzauer, and D. Käthner. Should My Car Drive as I do? What Kind of Driving Style Do Drivers Prefer for the Design of Automated Driving Functions? *Braunschweiger Symposium*, Vol. 10, No. 11, 2016, pp. 185–204.
19. Nagasaka, N., and M. Harada. Towards Safe, Smooth, and Stable Path Planning for On-Road Autonomous Driving under Uncertainty. *Proc., 2016 IEEE Conference on Intelligent Transportation Systems*, Rio de Janeiro, Brazil, 2016, pp. 798–801.
20. Gonzalez, D., V. Milanes, J. Perez, and F. Nashashibi. Speed Profile Generation based on Quintic Bézier Curves for Enhanced Passenger Comfort. *Proc., 2016 IEEE Conference on Intelligent Transportaton Systems*, Rio de Janeiro, Brazil, 2016, pp. 814–819.
21. Liu, P., A. Kurt, K. Redmill, and U. Ozguner. Classification of Highway Lane Change Behavior to Detect

Dangerous Cut-in Maneuvers. *Proceedings of Transportation Research Board*, 2016.

22. Omae, M., T. Fujioka, N. Hashimoto, and H. Shimizu. The Application of RTK-GPS and Steer-by-wire Technology to the Automatic Driving of Vehicles and an Evaluation of Driver Behavior. *Proceedings of IATSS Research*, Vol. 30, No. 2, pp. 29–38.
23. Ishibashi, M., M. Okuwa, S. Doi, and M. Akamatsu. Indices for Characterizing Driving Style and their Relevance to Car Following Behavior. *Proc., SICE 2007 Annual Conference*, Takamatsu City, Japan, 2007, pp. 1132–1137.
24. Bézier, P. Procédé de définition numérique des courbes et surfaces non mathématiques, système UNISURF, *Automatismes*, Vol. 13, 1968, pp. 189–196.
25. SAE International. *Automated Driving—Levels of Driving Automation are Defined in New SAE International Standard J3016*, SAE International. https://www.sae.org/misc/pdfs/automated_driving.pdf. Accessed July 15, 2017.
26. Welch, B. L. The Generalization of "Student's" Problem When Several Different Population Variances are Involved. *Biometrika*, Vol. 34, No. 1–2, 1947, pp. 28–35.

The Standing Committee on Vehicle-Highway Automation (AHB30) peer-reviewed this paper (18-00567).

Article

TRR
JOURNAL OF THE TRANSPORTATION RESEARCH BOARD

Transportation Research Record
2018, Vol. 2672(22) 10–23

DOI: 10.1177/0361198118799034
journals.sagepub.com/home/trr
SAGE

Lessons Learned from the Real-World Deployment of a Connected Vehicle Testbed

Mashrur Chowdhury[1], Mizanur Rahman[1], Anjan Rayamajhi[2], Sakib Mahmud Khan[1], Mhafuzul Islam[1], Zadid Khan[1], and James Martin[3]

Abstract

The connected vehicle (CV) system promises unprecedented safety, mobility, environmental, economic, and social benefits, which can be unlocked using the enormous amount of data shared between vehicles and infrastructure (e.g., traffic signals, centers). Real-world CV deployments, including pilot deployments, help solve technical issues and observe potential benefits, both of which support the broader adoption of the CV system. This study focused on the Clemson University Connected Vehicle Testbed (CU-CVT) with the goal of sharing the lessons learned from the CU-CVT deployment. The motivation of this study was to enhance early CV deployments with the objective of depicting the lessons learned from the CU-CVT testbed, which includes unique features to support multiple CV applications running simultaneously. The lessons learned in the CU-CVT testbed are described at three different levels: (i) the development of system architecture and prototyping in a controlled environment, (ii) the deployment of the CU-CVT testbed, and (iii) the validation of the CV application experiments in the CU-CVT. Field experiments with CV applications validated the functionalities needed for running multiple diverse CV applications simultaneously using heterogeneous wireless networking, and meeting real-time and non-real-time application requirements. The unique deployment experiences, related to heterogeneous wireless networks, real-time data aggregation, data dissemination and processing using a broker system, and data archiving with big data management tools, gained from the CU-CVT testbed, could be used to advance CV research and guide public and private agencies for the deployment of CVs in the real world.

Connected vehicle technologies (CVT) offer unprecedented safety, mobility, environmental, economic, and social benefits, which can be unlocked using the enormous amount of data shared between vehicles and infrastructure (e.g., traffic signals, centers) (*1*). The real-world connected vehicle (CV) pilot deployments will materialize these promised benefits of the CVT, which are needed to make informed investment and deployment decisions for large-scale deployments by state or local public agencies. The United States Department of Transportation (USDOT) announced a proposed law to install vehicle-to-vehicle (V2V) and vehicle-to-infrastructure (V2I) communications using dedicated short-range communication (DSRC) for all new lightweight vehicles to support different CV applications (i.e., safety, mobility and environmental applications) (*2*). From the perspective of this federal agency, state transportation agencies should deploy CVT and use the CV-generated data from the deployment sites to support state-wide planning and policies (*3*).

Although the USDOT has recently funded three pilot deployment sites in New York, Wyoming, and Florida to evaluate the efficacy of CV applications (*4*), the implementation guidelines for state or local agencies are still in the nascent stages of development. Initiated in Ann Arbor, Michigan in 2010, the CV Safety Pilot Program involved an analysis of the safety benefits of 2800 CVs at 29 infrastructure sites (*5*). Although the deployment program was successful in terms of elucidating the safety benefits of CVs, studies related to this program do not provide detailed guidance for the cases where multiple CV applications must be operated simultaneously that would require a heterogeneous wireless networking

[1]Glenn Department of Civil Engineering, Clemson University, Clemson, SC
[2]Department of Electrical and Computer Engineering, Clemson University, Clemson, SC
[3]School of Computing, Clemson University, Clemson, SC

Corresponding Author:
Address correspondence to Mizanur Rahman: mdr@clemson.edu

strategy for selecting specific communication options based on the corresponding application requirements in terms of data delivery delay and coverage range, data analytics capabilities for real-time data streaming, and data archiving strategies using big data management tools and infrastructure. Such deployment cases are required so that state public agencies may capitalize on the benefits from multiple CV applications, and provide academic and research institutions the means to develop innovative CV applications without direct involvement of the on-site CV deployment.

The Clemson University Connected Vehicle Testbed (CU-CVT) is a CV deployment site that includes unique heterogeneous wireless communication, data analytics for real-time data streaming, and data archival capabilities to empower the users with capabilities to operate multiple and diverse CV applications simultaneously. This paper discusses the case study of the deployment of the CU-CVT with the goal of sharing the lessons learned from the CU-CVT deployment, which can be leveraged by other public agencies, and academic or research institutions. Thus, the motivation of this study is to promote early large-scale CV deployments by disseminating findings in the deployment and operation of unique CU-CVT features.

The remainder of this paper is structured as follows. The next section describes existing connected vehicle testbeds and their limitations, and is followed by a section that presents the CU-CVT and the unique features of this testbed. Lessons learned from this testbed are then presented. The final section provides a concluding discussion.

Existing CV Testbeds

The USDOT leads the pilot deployment effort for CV testbeds in the United States through the ITS Strategic Plan (*6*). Southeast Michigan was one of the sites for Connected Vehicle Testbed, which is sponsored by the USDOT, and is available to DSRC product developers and CV application developers to test any CV applications. Testbed development began in 2009, and the project was completed in August 2017. Subsequently, the USDOT extended support for the sharing of CV-related data to develop a common technical platform for affiliated partner CV testbeds, such as the California and New York testbeds (*7*). Key CV technologies are related to the broadcasting of signal phase and timing (SPaT) data and intersection geometry (i.e., MAP) data, and CV devices (i.e., vehicle awareness devices, aftermarket safety devices, and road side units (RSUs)), and the USDOT preliminary Security Credentials Management System (SCMS) (*8*). The two ongoing projects are the "5.9 GHz Dedicated Short-Range Communication Vehicle-based Road and Weather Condition Application: Phase 2" and "Basic Infrastructure Message Development and Standards Support". The purpose of the first project is to develop and test the "weather condition information" transmission from CVs (i.e., public agency vehicles with connectivity used as CVs) to a central server via RSUs using 5.9 GHz DSRC communication. Public agency maintenance personnel use this stored information for future maintenance purposes.

In 2012, the University of Michigan Transportation Research Institute and the USDOT started a new project to reduce roadway crashes using CV technology (known as Safety Pilot Model Deployment (SPMD) [*9*]). The total investment for this three-year SPMD project (2015–2018) was $30 million, which incorporated over 2,800 vehicles and 73 lane-miles of northwest Ann Arbor, Michigan. After successful implementation of SPMD, the Ann Arbor testbed was extended from northeast Ann Arbor to the City of Ann Arbor. The new testbed covers 27 sq. mile city of Ann Arbor including 45 street locations and 12 freeway sites. Every year 1,500 vehicles are to be added in this deployment with the goal of having 5,000 vehicles on the road by 2018. The main goals of this testbed are to ensure that (i) the model deployment will be transferrable to an early operational deployment; (ii) the test facilities will continue to operate and maintain a robust and high-quality test environment; and (iii) the testbed can become an economically sustainable environment from a federally funded program. It will become the world's largest CV infrastructure deployment, and it is called the Ann Arbor Connected Vehicle Test Environment.

The USDOT awarded a total of $45 million to sites in New York, Wyoming, and Florida in September, 2016 to initiate design, deployment and validate the CV Pilot Deployment Program (*10*). The primary focus of the New York City (NYC) pilot deployment in New York is to improve vehicle and pedestrian safety (*10*, *11*). This pilot deployment includes 280 RSUs and 10,000 vehicles (cabs and buses), and will be used to evaluate CV safety challenges and benefits in an urban environment. Furthermore, they will evaluate pedestrian safety applications using 100 pedestrian-DSRC units. Multiple locations on Interstate 80 were chosen as the Wyoming CV pilot deployment site in another project (*12*). This project deploys DSRC on-board units in 400 CVs (mainly snowplows and heavy vehicles) and provides safety alerts and travel guidance to the commercial CVs. In Florida, the Tampa-Hillsborough Expressway Authority (THEA) pilot deployment plans to implement a variety of V2V and V2I applications (*13*). The focus of implementing these applications is to reduce congestion, conflicts between vehicles, and entry of a vehicle from wrong direction at the Selmon Reversible Express Lanes exit.

Another purpose of the THEA pilot deployment involves using the CV technologies to improve the safety of pedestrians, increase efficiency of bus operations, and reduce conflicts between vehicles and pedestrians. This deployment uses DSRC as the wireless data transmission medium. Through this CV pilot program, 1,600 cars, ten buses, ten trolleys, 500 pedestrians with smartphone applications, and 40 RSUs along city streets will be deployed in Tampa, Florida.

The first public CV testbed in the United States was developed in 2005 at California on El Camino Real (State Route 82) in Palo Alto between Stanford Ave at the north end and W Charleston Ave (*14*). There are 11 consecutive intersections exists along this corridor. More than 50,000 vehicles travel this corridor between San Francisco and San Jose daily. The partners of this testbed are Caltrans, the Metropolitan Transportation Commission and the California PATH program at UC Berkeley. In 2013, CV standards were implemented in the previously developed California CV testbed with the support of USDOT. The focus of this testbed is deploying multimodal intelligent traffic signal systems and environmentally friendly driving applications. The primary objective is to improve overall arterial network performance utilizing traffic signal priority for transit and freight, preemption for emergency vehicles, and pedestrian safety applications. Besides, Virginia has two CV testbeds (*15*), the first located in Southwest Virginia in Blacksburg Virginia at the Virginia Smart Road, and the second along Route 460 and Fairfax County in Northern Virginia along I-66 and on the parallel Routes 29 and 50. The Connected Vehicle/Infrastructure University Transportation Center (CVI-UTC) selected these two locations for their CV research based on congestion, high crash rates, and air quality non-attainment. Fifty roadside units were deployed along the roadways of both, which also include different instrumented vehicles (e.g., instrumented automobiles, motorcycles, a motor coach, and a semi-truck).

In existing CV testbeds, CVs are deployed to test CV safety applications using DSRC. Although DSRC is a very low-latency wireless communication medium, it is limited in terms of communication range (~900 ft (300 m)) (*16*). Thus, DSRC alone cannot support the breadth of CV applications (i.e., mobility and environmental applications) outlined by the connected vehicle reference implementation architecture (*17*). Other wireless technologies, such as long-term evolution (LTE), WiMAX, and Wi-Fi can support DSRC communications to increase network availability and coverage area, and fulfill data rate requirements at peak period. Thus, it is necessary to use heterogeneous wireless technology in which communication options can be chosen based on the requirements identified by CV applications, and their range, data delivery delay, and throughput. Another limitation of existing testbeds is the missing data analytics platform for real-time data aggregation, processing, and archiving for multiple CV applications running simultaneously. Designing a suitable data delivery system for CVs depends on the real-time data streaming from different devices without duplication and supplying the aggregated data depending on the application demand (*18*). Appropriate computational resources are also necessary to support multiple and computationally expensive CV applications.

Unique Features of CU-CVT Testbed

The CU-CVT is developed in a hierarchical architecture. In a CV environment, different types of applications run at different levels (e.g., vehicle level, roadside infrastructure level, system level (i.e., backend server or cloud)) (*19*). A layered architecture has proven effective in managing such distributed applications by decoupling the dependency and distributing the computation. The overview of the layered architecture in CU-CVT is shown in Figure 1. The CU-CVT layered architecture has a (i) Mobile Edge, (ii) Fixed Edge, and (iii) System Edge. The Mobile Edges will reside in Layer-1 of the architecture. Each Mobile Edge or each mobile entity (e.g., vehicles, pedestrians) is equipped with a DSRC, 3GPP/LTE Cellular or Wi-Fi-enabled device to collect information related to the Basic Safety Messages (BSMs) such as time stamp, car ID, latitude, longitude and speed. To support different types of application with different levels of computation and memory requirements, DSRC devices are coupled with a computation unit capable of larger memory management, multiprocessor operations, and user-friendly application development. The computation unit also permits the integration of other communication mediums, such as Bluetooth, Wi-Fi, and LTE. The Fixed Edge acts as an entry point to the infrastructure with a very low-latency, single-hop DSRC connection with the Mobile Edge. Residing in Layer-2 of the system, the Fixed Edges communicate with components in Layer-1 and Layer-3. Fixed Edges can be extended to support a video camera and other sensing devices (e.g., weather sensors). Fixed Edges also contain a computational platform for the same reason as the Mobile Edge devices, along with a DSRC-enabled device. Fixed Edge devices are connected to Layer-3 with a high-bandwidth, high-throughput, low-latency and highly reliable communication medium (e.g., fiber optics, wired Ethernet) to act as backhaul support.

The System Edge is situated at the top of the layered architecture, Layer-3. The task of the System Edge is to monitor the overall system to balance and optimize system operations and provide services to different nodes

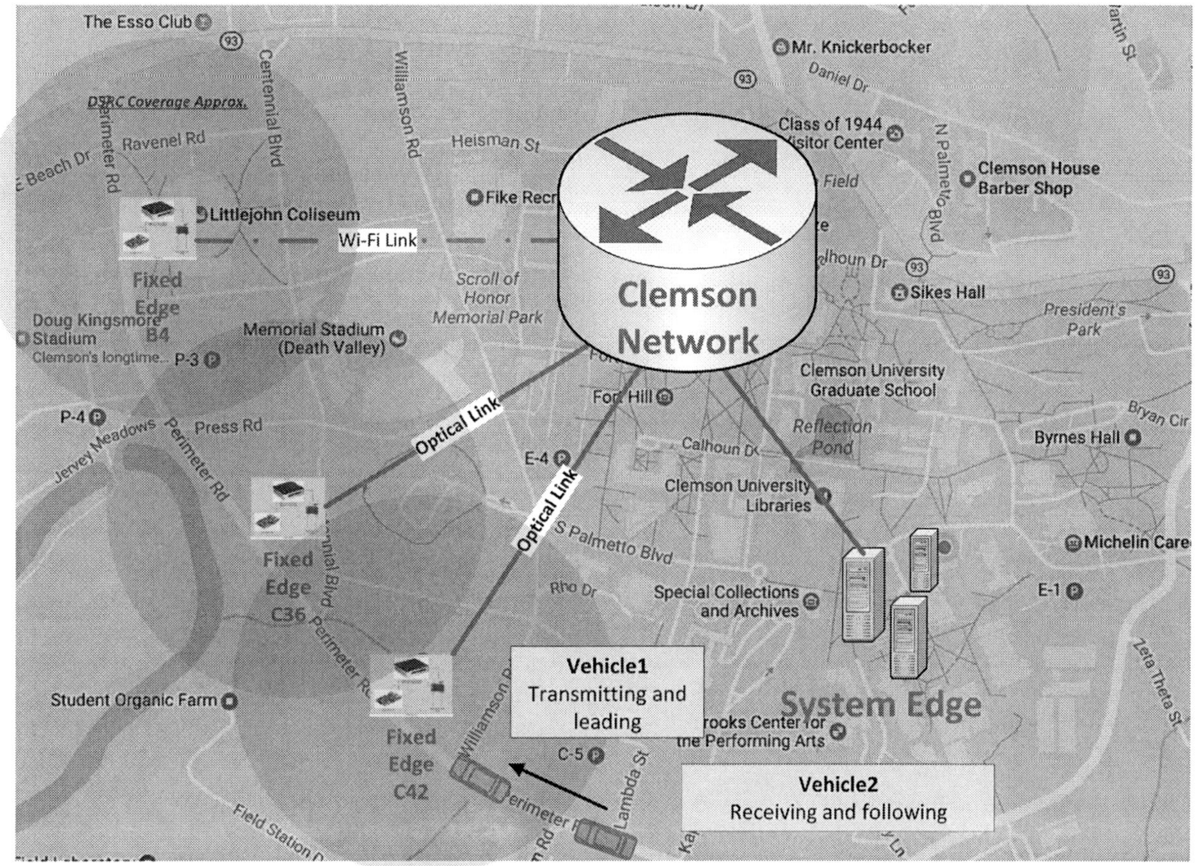

Figure 1. Clemson University Connected Vehicle Testbed (CU-CVT).

related to connectivity, data storage, application priority, and dissemination of system-wide messages, and to act as a secure portal for external agencies and research interests. One of the main responsibilities of the System Edge is to provide resource optimization and management of communication network resources in heterogeneous wireless networks. Conceptually, the System Edge can be in a cloud or cluster of computers. The System Edge used here supports the various services mentioned above.

CU-CVT system and CV applications are closely intertwined with wireless access in vehicular environments (WAVE) message sets stated in Society of Automotive Engineers (SAE) J2735 (*20*) and IEEE 1609.3 standards (*21*). CU-CVT supports standard WAVE applications provided by the equipment vendors, which are a set of safety applications including collision avoidance. The CU-CVT testbed provides users and developers the ability to operate a variety of applications and services that are interoperable with standard DSRC WAVE. The applications comply with DSRC standards in terms of transmission channel, frequency, and power (*22*). Any other DSRC radios, complying with the 1609 WAVE standards and licensed to operate in the U.S., will be able to work with these applications utilizing standard DSRC channels. The goal is to make the CU-CVT system highly interoperable with any other CV applications that are compatible with WAVE multichannel standards (i.e., IEEE 1609.4) (*22*). In addition, the testbed and the hierarchical architecture provide a solid framework for implementing security-enforcing services. The CU-CVT testbed and CV applications are designed to use the vendor-specific security keys that are embedded into each licensed DSRC unit. The applications also provide methods to enable and disable such privacy enforcements. With the hierarchical structure of the testbed, implementation of SCMS (*23*) or other forms of security and privacy methodologies can be easily implemented. In addition to the security and privacy platform available on DSRC networks, CU-CVT testbed has a virtual private network (VPN) on top of Clemson University campus network. The private network being used for the testbed provides an impenetrable backbone for data communication between Fixed Edges and a System Edge. Collectively, the VPN and the use of secret private keys exchanged before initiating any communication satisfy the privacy requirements for the testbed and the application data.

Lessons Learned: CU-CVT Deployment

This section presents the lessons learned at different levels of the CU-CVT deployment: (i) system design and prototyping; (ii) system deployment; and (iii) system validation.

System Design and Prototyping

In the process of deploying the real-world CV testbed, the first step involves defining the system architecture, selecting an appropriate communication medium for use among the CV components and software components, and evaluating and justifying these selections by developing a prototype system. In this section, the lessons learned from the system design architecture and prototyping in a controlled environment are detailed.

System Design Architecture. The focus of this sub-section is to define the rationale for selecting a layered design architecture, heterogeneous wireless communication medium, data analytics using real-time data streaming, and data archiving with big data management tools in the process of concept development of the CU-CVT.

Layered Architecture. The key motivations for choosing the layered architecture of the testbed are to reduce data delivery delay and data loss rate, and increase throughput between diffferent edges of the CU-CVT testbed (*19*). DSRC provides a low-latency, high-throughput communication link from a Mobile Edge to a Fixed Edge and many applications, such as queue prediction, traffic data collection, and video monitoring, which may need computation at the Fixed Edge rather than sending the data to a System Edge. Thus, the overall effectiveness of the application is improved in terms of analyzing the data, which is a primary motivation for designing the layered architecture. Layered architecture can also support the integration of roadway sensors (e.g., video cameras, weather sensors, roadway traffic sensors) with the CVT deployment while reducing high bandwidth requirements. There are several advantages of this layered design architecture including proximity, intelligence, and scalability. Proximity of the edge provides the flexibility to communicate efficiently with the next immediate edge layer, and information distribution will be quicker than with a centralized system. The next immediate layer is defined as the closest edge layer in terms of the physical and logical sense. With the growing number of CVs, massive amounts of raw data will be collected and processed into usable formats for different CV applications. Thus, intelligence functions must be distributed efficiently between different layers to reduce the computational time for computationally expensive applications. This layered architecture can be easily scalable by changing the number of edges in different layers to meet the demand of computational resources.

Heterogeneous Wireless Communication (Het-Net). To ensure the reliability of data transfer for ITS applications, a combination of different wireless communication technologies such as LTE, Wi-Fi, and WiMAX has been used (*24*, *25*). However, establishing the proper communication networks is a challenge in a layered architecture with dynamic vehicular networks. Traditional network management is unsuitable for highly mobile vehicular networks because of its dependency on network access, and packet management and routing requirements. Het-Net allows selection between different wireless communication options based on their availability, reliability, and data delivery requirement of the CV applications (*26*, *27*). Based on the CV application requirements, Het-Net optimizes networking resources, which can provide a better solution to improve communication and content delivery issues with CVT.

Data Analytics using Real-Time Data Streaming. One of the primary challenges in layered architecture involves managing the data delivery system between different components. To mitigate this challenge, CU-CVT has adopted a publish–subscribe-based data delivery system among the connected components (*18*). A broker-based system, which is a publish–subscribe-based data delivery system, is a popular data delivery system in different areas that include social media, e-commerce, and manufacturing (*18*). In general, in this system, producers (e.g., sensors, RSUs) produce data, and consumers (e.g., CV applications, emergency management center, and traffic management center) consume data. Such frameworks include Kafka, RabbitMQ, ActiveMQ, WebsphereMQ, and Message Queuing Telemetry Transport (MQTT) (*18*).

Data Archiving using Big Data Management Tools. Along with the data delivery system, a data storage system is also necessary. A database is a piece of software and storage area where data can be stored that can be used for data analysis and reporting. In a CV environment, the different types of structured and unstructured data generated from different sources (e.g., on-board units, RSUs, sensors) must be aggregated and stored at

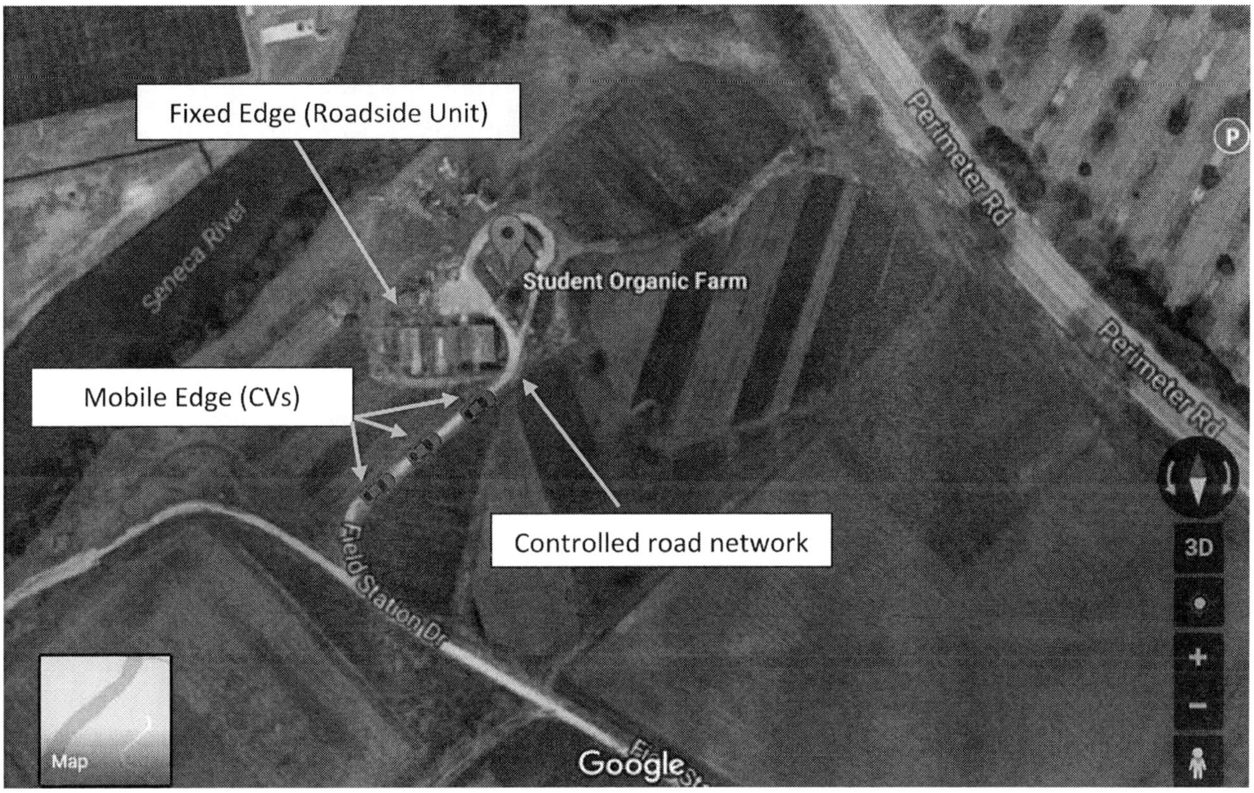

Figure 2. Organic farm prototype deployment in a controlled environment.

different edge levels of the layered architecture. Structured data can be stored in relational databases (structured query languages (SQL)), and non-structured data are suitable for storage in NoSQL databases (*28*). Oracle, MySQL, and SQL Servers are the most popular relational database systems used in industry, whereas MongoDB, CouchDB, and BigTable are the most popular NoSQL databases (*29*, *30*). Both have some advantages and disadvantages in terms of performance, usability, and scalability. The evaluation and selection of the most suitable data management tool for the CV environment is described in the sub-section "Data Archiving using Big Data Management Tools" of the System Deployment section.

System Prototype Development in a Controlled Environment Before the real-world deployment of the CV system, it is important to develop a prototype of the application and system architecture in a controlled environment, especially in a controlled roadway. Given that a focus of the CV applications is that of safety-critical issues, deploying any application without such testing may create a hazardous situation. The first phase of application or research development involves testing the process in a controlled lab environment where the operation of the application, logic, and errors are debugged and perfected. Therefore, a particular corridor at Clemson University was selected (i.e., the Student Organic Farm, as shown in Figure 2) as the real-world CV application test site. After the successful development and testing in a controlled environment, the application was deployed in the real world, which eventually involved CU-CVT.

System Deployment

In this section, the lessons learned from the CU-CVT deployment are described in detail.

Location Selection. Identifying the optimal locations of physical devices is a critical aspect of field deployment. This system was deployed on the main campus of Clemson University (Clemson, SC) as illustrated in Figure 3, with three Fixed Edges positioned along Perimeter Road. The location of the RSU depends on the surrounding environment, roadway geometry, opportunities for roadside mounting, and the purpose of the testbed. The specific range of functionality of the RSUs (owing to DSRC range limitation) requires placing them at intervals to ensure only a minimal range

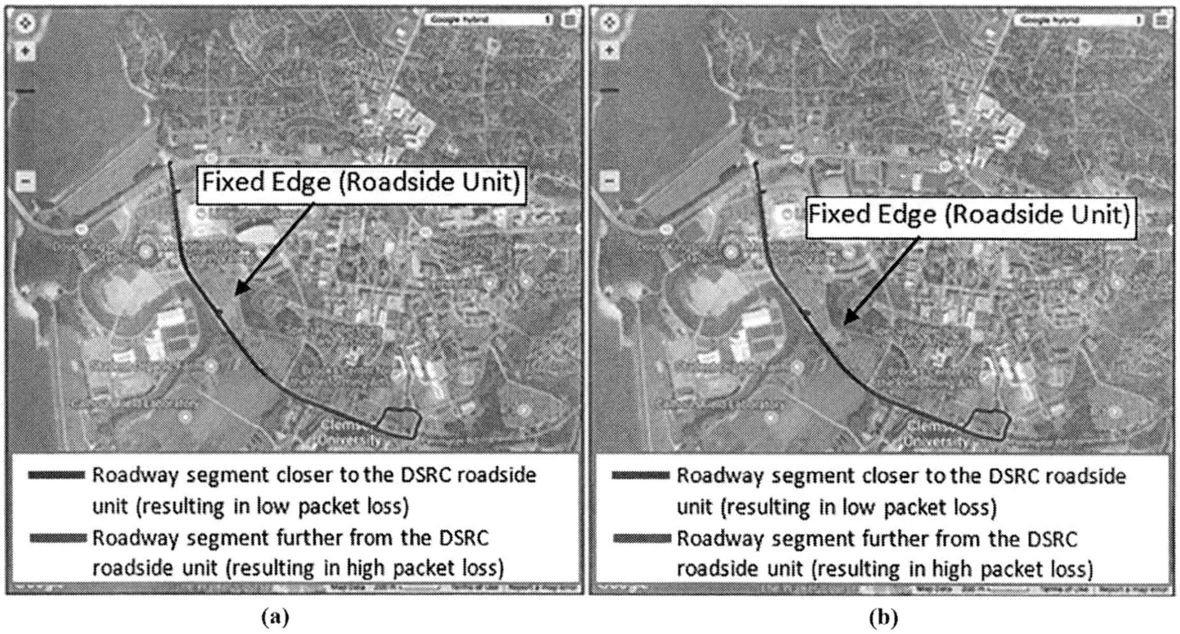

Figure 3. (*a*) Packet loss at C36 location, (*b*) Packet loss at C42 location.

overlap. The blue line in Figure 3 shows the range from the field test.

The effect of RSU location on packet loss at two locations (C36 and C42) on the Perimeter Road deployment is shown in Figure 3. The RSU device used is the same for both RSU locations. One RSU shows higher coverage (Figure 3*a*), whereas the other location (Figure 3*b*) shows less coverage because of the horizontal and vertical curve of the roadway. This is because of external factors such as interference and signal blockages. Thus, the location should be selected depending on the road way geometry and topography. The RSUs should be placed at locations with minimum interference from trees, houses, and other signal blockages. For the CU-CVT deployment, a corridor with multiple signals was chosen, and signalized intersections were used as the RSU locations. An intersection that allows the use of RSU for CV safety, and mobility applications related to intersections (e.g., queue warning, stop-sign gap assist) is considered to be a suitable location. Consecutive intersections have a certain spacing, allowing installation of one RSU at each intersection and covering the whole corridor. It also serves the purpose of covering multiple directions, which may be required for some CV applications (e.g., stop-sign gap assist). Finally, the close proximity of the RSU to the signal controller provides easy access to the controller, which may also be valuable in V2I applications.

System Compatibility and Integration. One of the key issues for real-world deployment is system compatibility. These deployment experiences showed that different vendors have different ways of implementing applications, services, communication, and data streaming application programming interface. The main problem experienced is that application development environment varies in different vendor devices. The testbed should be independent of vendor choice. As there is no specific standard for device configuration and it is the early phase of equipment production, one needs to write one's own code to make it compatible with the platform. Another approach involves developing a generalized CVT distributed application services package, such as ThinGs in a Fog (TGIF) (19), to provide services, such as geo-location, messaging, HetNet and WAVE services, for application developers.

Heterogeneous Wireless Communication. Het-Net allows the selection of the best available communication options based on their availability and reliability, as well as data delivery requirements of the corresponding CV applications. Figure 4 illustrates an example deployment scenario with three communication options, Wi-Fi, LTE, and DSRC, in which a wireless network will be selected based on the requirements of the CV application to be supported, coverage area and availability of the network, and handoff delay for switching from one wireless communication option to another wireless communication option. As shown in Figure 4, CVs (Mobile Edges) use

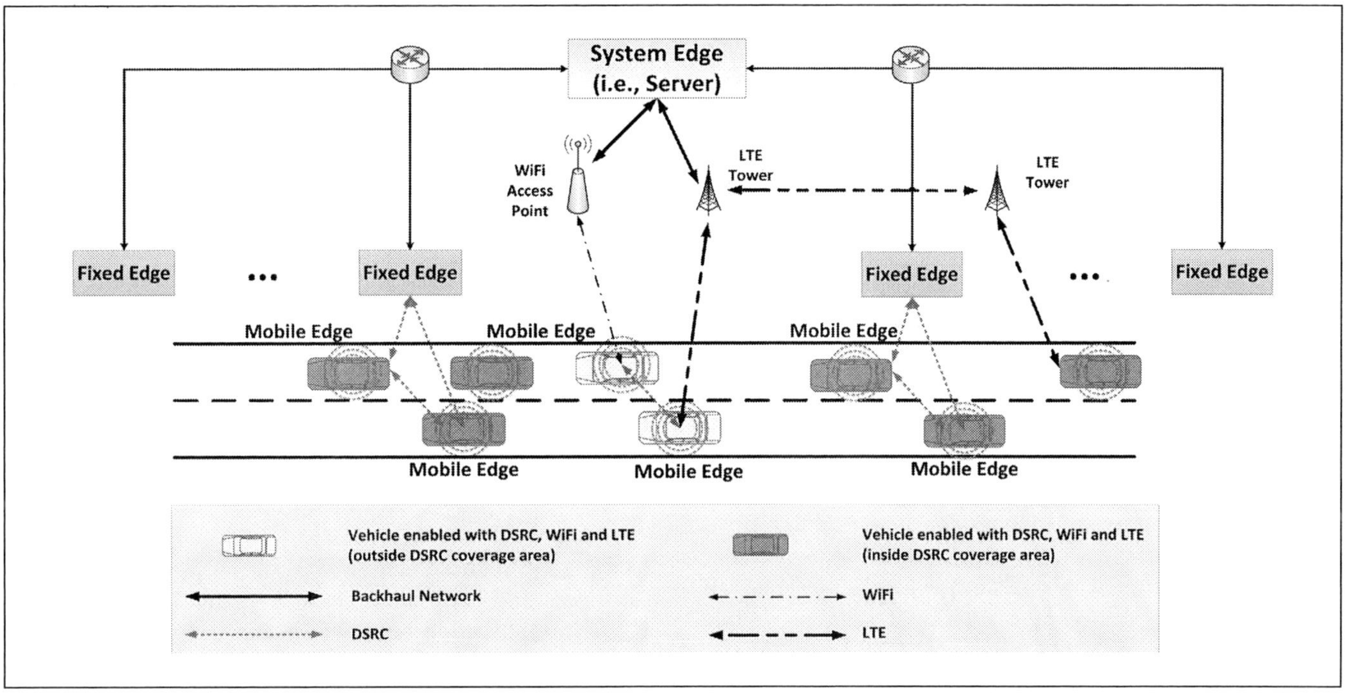

Figure 4. Heterogeneous wireless communication scenarios.

DSRC network to send data to the Fixed Edges while they are within the DSRC coverage area of a Fixed Edge (RSU). From the Fixed Edge, data collected from the CVs will be transferred to the System Edge using backhaul network (either through wired, Wi-Fi, or any long-range wireless communication network). On the other hand, CVs use LTE wireless network to transfer data directly to the System Edge while they are outside the DSRC coverage area of a Fixed Edge (RSU). CVs can also use Wi-Fi wireless network to transfer data directly to the System Edge if Wi-Fi network is available. This study has deployed two Het-Net instances (instance 1: switching of networks between LTE and Wi-Fi; instance 2: switching of networks between LTE and DSRC) based on the availability of the communication options and subject to the CV application requirements. An application layer handoff, which involves switching or "handoff" from one network to another, is an elemental approach for managing heterogeneous networks. The application layer handoff works with no change to hardware and software of the Mobile Edge (on-board unit) and Fixed Edge (RSU). A hard-handoff is performed when one wireless network is switching to another wireless network to implement Het-Net services. The hard-handoff can be defined as when a wireless communication network connection is broken before connecting with another network.

Data Analytics with Real-Time Data Streaming. The messaging service is equipped with a broker or the ability to publish/subscribe to a broker. For the CU-CVT deployment, the MQTT broker was used (*30*). MQTT is a lightweight open source publish/subscribe messaging transport protocol, and it is easy to implement. MQTT is superior to the hypertext transfer protocol (HTTP) messaging service, because it provides better throughput, has lower power consumption, lower network bandwidth utilization, and lower latency. The publish/subscribe protocol makes the communication between different edges efficient as it enables a large number of mobile clients (e.g., Mobile Edge) reliably communicate with Fixed Edges or a server (e.g., System Edge), and vice versa in an edge centric connected vehicle testbed like the CU-CVT. Publish/subscribe decouples a client (message sender) from another client (message receiver) by making use of a unique set of strings for each type of message called Message Title. The MQTT broker filters all the received messages and distributes those messages accordingly. MQTT uses subject-based filtering of messages. CV applications are dynamic in nature, and most applications require efficient real-time data streaming and data distribution between different edges of the CU-CVT. Thus, data exchange in V2V and V2I communication is a major challenge. Moreover, CV applications operate in a constrained environment, where there are often limitations in bandwidth, computation, and storage. Given the requirements of a lightweight broker-based communication method, MQTT is a good choice for the field deployment of CV applications.

Data Archiving with Big Data Management Tools. Data storage is an essential part of CV applications. Many CV applications have to deal with big data, where there is a large volume of streaming data and a large variety of data (i.e., real-time location, vehicle speed, position, acceleration). The choice of database for archiving data should be uniform across all layers of field deployment. For the CU-CVT deployment, MongoDB was used as the uniform data storage platform for CV applications. As CV applications will need to deal with huge amounts of data, a non-relational database is a better choice than a relational one. NoSQL databases offer advantages over SQL databases in several aspects, such as scalability and data diversity (i.e., different types of data). Among many NoSQL databases, MongoDB was chosen for the CU-CVT deployment because it has many advantages over other NoSQL databases. MongoDB is a popular open source NoSQL database tool for big data management. Being open source is a major advantage. MongoDB offers the query ability of a relational database while maintaining the features of a NoSQL database. Moreover, if the data are location based, MongoDB is handy because it has some built-in spatial functions. MongoDB stores data using the very popular JavaScript Object Notation (JSON) format, which is compatible with all programming platforms. Different CV applications require different data storage requirements and different data structures. Therefore, a fast, scalable, and flexible data storage platform is needed that can store any type of data, of any size, at any time while the application is running. Moreover, different applications may operate on different platforms. A universal data storage platform is needed that will perform effectively across all CV applications, and MongoDB is the solution.

Inter-Agency Cooperation. The challenge of developing a testbed for field implementation requires collaboration with multiple stakeholders. For the CU-CVT deployment, significant collaboration was required between the researchers and the Clemson University Information Technology (IT) department. The research team often works closely with the IT department of the university, as the campus Wi-Fi network (Eduroam) and optical fiber is used for backhaul support. To install a RSU at the testbed, a light pole close to the traffic signal was chosen. There was regular communication with the Clemson University campus physical planning department for the bucket truck to install the RSU at the Perimeter Road (where the CV-CVT is located) light pole. The light pole was selected so that there are optical fiber services from the pole to the backend resources. Without diligent support from the Clemson University campus physical planning and IT departments, the proper deployment of the testbed would have been impossible. The use of a state-owned road requires permission from the state Department of Transportation (state DOT) to set up the roadside RSUs and secure any access to the state DOT infrastructure, specifically the DOT's communication network and traffic control devices and equipment.

System Validation in the CU-CVT

The lessons learned from system validation include validation experiences on DSRC system performance ascertainment, heterogeneous wireless network, data analytics with real-time data streaming, and system integration.

DSRC System Performance Ascertainment. The CV testbed is deployed on a geographically spread out network, including dynamic scenarios ranging from high-speed less congested scenarios to slow-speed highly congested scenarios. For these scenarios, it becomes very important to understand how the system performs to provide reliable communication between different Mobile Edges and Fixed Edges. The coverage range obtained for the three Fixed Edges in the CU-CVT is shown in Figure 5, with the received signal strength indicator value measured by the Fixed Edges. This coverage of the DSRC range provides an observation on network availability and allows for strategic planning for heterogeneous network management and switching between different RSUs.

Heterogeneous Wireless Communication. In this study, two Het-Net instances were evaluated (instance 1: switching of networks between LTE and Wi-Fi; instance 2: switching of networks between LTE and DSRC) in terms of handoff delay. The average delay recorded between successive packet arrivals is used to estimate the handoff delay during switching from one network to another network. UDP data transfer protocol is used to send data packets from a CV to the roadside infrastructure (i.e., Fixed Edge) or server (i.e., System Edge). In the first instance, Wi-Fi and LTE wireless Het-Net service is implemented for transferring data from vehicles to a System Edge (i.e., Server). For receiving data packets through Wi-Fi, the 802.11 link must be active and a CV must be within the coverage area of the Wi-Fi access point. The hard-handoff delays during switching from Wi-Fi to LTE and switching from LTE to Wi-Fi are 3.61 and 3.10 seconds, respectively.

In the second instance, LTE and DSRC wireless Het-Net service is implemented for transferring data from CVs to the roadside infrastructure or server. A vehicle equipped with DSRC radio (i.e., Mobile Edge or CV) can send data to Fixed Edge (i.e., RSU) when the CV is inside the DSRC coverage area of the Fixed Edge. Before entering the coverage area of the Fixed Edge, the LTE wireless network option of the CV is always active. As

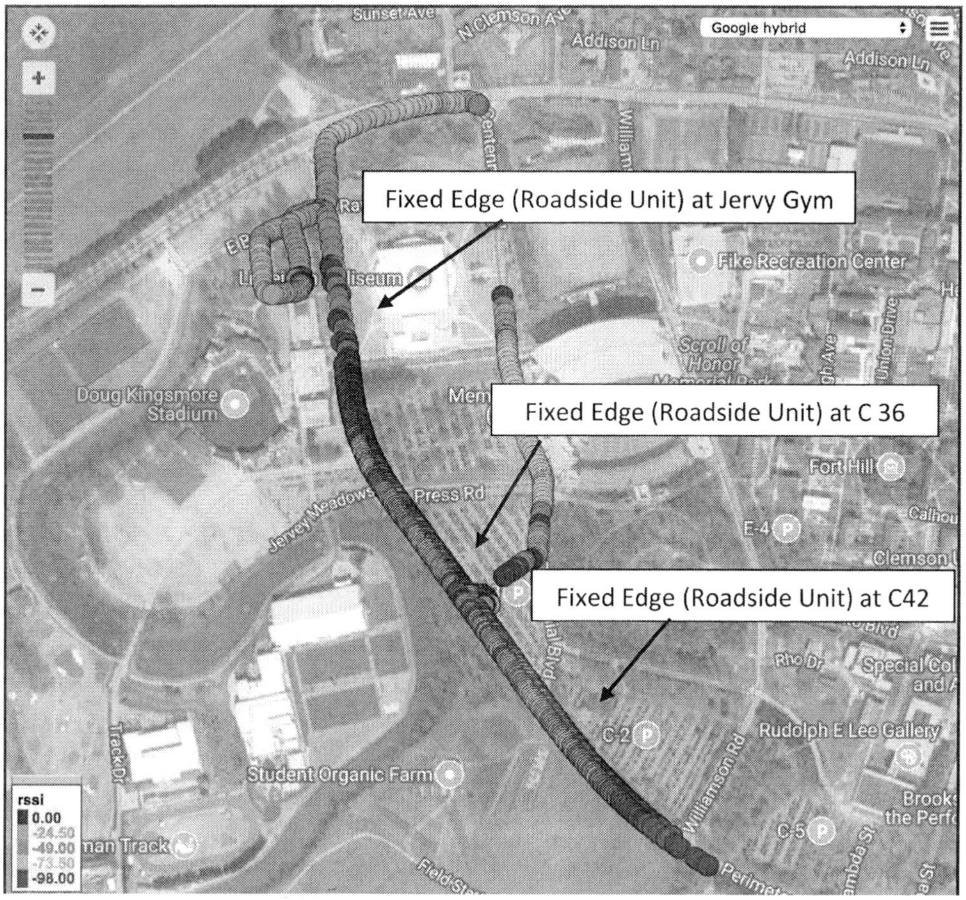

Figure 5. DSRC network coverage of the CU-CVT testbed.

the CV is equipped with DSRC communication capabilities, it receives the beacon-like message when it enters the coverage area of the Fixed Edge with DSRC, and an application inside DSRC radio of a CV initiates the handoff process. The application executes a hard-handoff while switching from LTE to DSRC. The opposite handoff process (switching from DSRC to LTE) is executed when the CV leaves the DSRC coverage area of the Fixed Edge and receives the beacon-like messages from the LTE wireless network. The handoff delays during switching from DSRC to LTE and switching from LTE to DSRC are 0.23 and 0.03 seconds, respectively, which are much smaller than the corresponding handoff delays between Wi-Fi and LTE.

Data Analytics with Real-Time Data Streaming. Raw data (e.g., time stamp, car ID, latitude, longitude and speed), known as BSMs, are collected from each CV. The campus Fixed Edge used Wi-Fi to communicate with the backhaul server (System Edge). Vehicles equipped with CV on-board equipment (i.e., Mobile Edge) could thus collect real-time data from other CVs as those vehicles traveling within their DSRC range. All Mobile Edge device-equipped vehicles send data to the RSU using DSRC if these vehicles are within the DSRC coverage area of a Fixed Edge. Optical fiber is used to connect with the System Edge with the Fixed Edge when real-time communication is required. A database system is hosted using MongoDB to store data at the Fixed Edge and at the System Edge. After collecting raw data from the DSRC on-board unit, the DSRC-RSU will publish raw data to the MQTT broker system, and later the raw data processing unit subscribes the MQTT broker to consume the required data. The MQTT broker, the MongoDB database, and other computational processes reside in an edge-computing device. The processed data are stored for a short period at the MongoDB database to run real-time CV applications in the Fixed Edge. On the other hand, in the System Edge, a MongoDB database is used to store data for a longer period so that other CV applications can use the stored data.

Table 1. Collision Avoidance Results for CV-2

	Safety distance		Message delivery latency	
Vehicle speed (mph)	Average safety distance of three test runs between CV-1 and CV-2 (when 1st warning message packet received) (ft)	Minimum safety distance requirement (considering 11.2 ft/s^2 deceleration rate) (ft)	DSRC average message delivery latency (ms)	Latency requirements for CV safety applications (ms) (*32*)
25–30	156	38	2.4	200
35–40	201	118	1.8	
45–50	316	240	1.7	

System Integration and Validation: A Case Study with Collision Avoidance and Queue Detection Application. To verify the accurate integration of the hardware and software components with reliable communication systems to ensure the requisite safety and mobility CV benefits, the overall system integration was validated by testing the collision avoidance and queue detection applications.

Collision Avoidance Application. From the field tests with the CU-CVT, the research team recorded the time required to deliver emergency safety messages from a vehicle that stopped suddenly (CV-1 or Mobile Edge 1) to the immediate following vehicle (CV-2 or Mobile Edge 2) using DSRC. Three different ranges of vehicle speeds (i.e., 25 -30 mph, 35 -40 mph, and 45 -50 mph) were used for the field tests. Table 1 provides the average safety distance between CV-1 and CV-2 in three test runs with three different ranges of vehicle speeds (i.e., 25–30 mph, 35–40 mph, and 45–50 mph). Also calculated was the minimum safety distance requirement for CV-2 (Table 1, Column 3) when CV-2 applied the brake after receiving the first warning message, with a deceleration rate of 11.2 ft/s^2 (*31*). The motivation of this field test is to determine if measured DSRC message delivery latency can fulfill the requirement of a collision avoidance application after receiving the first warning message. As shown in Table 1, the minimum safety distance required is less than the average safety distance available between CV-1 and CV-2 at all three ranges of vehicle speeds considering maximum speed of the range. In addition, the observed average message delivery latency is significantly less than the latency requirements for CV safety applications (*32*). The receiving time of the safety message in the vehicles downstream depends on several factors, such as vehicle movement patterns, vehicle speeds, and any interference or noise in the environment. Thus, the distance to the upstream vehicle at the time of receiving the safety message varied for different tests. Simultaneously, the exchange of safety messages between CV-1 and CV-2 served as a prompt for the following three events:

1. Using LTE, the Het-Net system in CU-CVT delivered the collision-warning message from CV-1 and CV-2 to another vehicle (CV-3) which was beyond the DSRC range;
2. CV-1 sent BSMs to the nearest Fixed Edge via DSRC as CV-1 was within its DSRC range;
3. The Fixed Edge sent the BSMs to the System Edge layer using the Wi-Fi communication option available in CU-CVT.

In previous work (*26*), the research team proved that the latency for LTE wireless communication is above the required latency for CV safety applications but within the latency requirement to receive possible upstream incident warning. This discretion in latency proves the suitability of Het-Net for CV applications enabling multiple wireless networking options based on the latency requirements of the CV applications. A database system was hosted using MongoDB to store data at the Fixed Edge and at the System Edge. After collecting BSMs from the DSRC-enabled CV-1, the DSRC-enabled RSU at the Fixed Edge will publish BSMs to the MQTT broker system, and MongoDB archives BSMs from the MQTT broker. BSMs are also archived in the MongoDB database at the System Edge for future use.

Traffic Queue Detection Application. A queue ahead of a traveling vehicle is one of the major causes of rear-end collisions that can also disrupt traffic throughput by introducing shockwaves into the upstream traffic flow (*33*). Daily recurring congestion, work zones, incidents, and/or weather conditions are also causatives of traffic queues (*33*). Rear-end collisions and the impact of resulting traffic flow shockwaves can be significantly reduced through the rapid detection of a traffic queue, the prompt formulation of an appropriate response message for approaching vehicles, and the dissemination of queue information to the approaching vehicles in real time. A threshold-based queue detection algorithm was developed by USDOT using V2V and V2I communication utilizing DSRC (*33*). Typically, the threshold-based algorithm uses average speed of the vehicle (i.e., less than 5 mph) as well as the average separation distance (i.e., less than 20 ft) within DSRC communication range. Each CV in the DSRC range can broadcast a series of BSMs every 1/10th of a second. The research team implemented the threshold-based queue detection algorithm and evaluated queue detection application accuracy in the CU-CVT testbed. The queue detection application was used to effectively showcase the usage of different features of the testbed for developing and deploying CV applications. Both the average speed and average separation distance were used to detect a traffic queue in the testbed.

An application running in the Mobile Edge broadcasts vehicular information (i.e., time stamp, car ID, latitude, longitude and speed) to a Fixed Edge within the communication range of the DSRC. Using a python script at Fixed Edges, the separation distance between two consecutive CVs was calculated using the latitude and longitude of each vehicle. The speed and separation distance are also averaged of all three CVs for each second. Using the average speed and average separation distance, the queue was detected by running the developed threshold-based algorithm at Fixed Edges when three CVs stopped at the traffic signal in front of the Jervy Gym location. The accuracy of the threshold-based queue detection method is 81% for three test runs, and the threshold-based algorithm was run 167 times (every 1 second) for traffic queue detection. To calculate the accuracy of the threshold-based algorithm, the threshold-based queue prediction algorithm was compare with field-collected video camera data, which represents the ground truth data. As the accuracy of the algorithm depends on average separation distance between vehicles, the reason for lower accuracy is a higher separation distance between the CVs higher than the threshold because of the presence of non-connected vehicles between CVs. The backhaul network consisting of a Wi-Fi network is used to connect the Fixed Edges with the System Edge when real-time communication is required. Big data software, MongoDB, was used to store data at both the Fixed Edge and at the System Edge. This effort was used to demonstrate the suitability of the testbed for deploying both safety and mobility CV applications.

Conclusions

CU-CVT has several unique features, such as layered architecture, heterogeneous wireless communication medium, data analytics using real-time data streaming, and data archiving with big data management tools. The layered architecture is validated through a system design process. As the layered architecture is easily scalable depending on the deployment area and penetration level of CVs, the CU-CVT design is applicable for future large-scale deployment. In addition, the developed heterogeneous wireless technique can be easily implemented in many other CV testbeds. The selection between different wireless communication options (e.g., Wi-Fi, LTE) is based on the feasibility, accessibility, and data delivery requirements of the application. The use of this heterogeneous network technology also makes the expansion of traffic data collection for traffic management possible in terms of the data collection coverage area in a CV environment. The testbed and supporting software application programming interfaces are designed to make the application developed for the testbed independent of both different network technologies and networking devices. The testbed is also easily reconfigurable for forthcoming networking technologies such as 5G. The research team's goal has been to develop applications and systems that can be easily integrated to any platform and systems, and that support any new networking technologies. The motivation has been to develop a connectivity strategy in which the applications can be developed independently of networking and other CV device attributes. These features make the CU-CVT testbed and application development platform reach a longer life cycle in terms of future CVT research, transferability of the work, and supporting current and future CVT technology development.

Lessons learned were identified at three different levels: (i) the development of design architecture and prototyping in a controlled environment, (ii) the deployment of the CU-CVT testbed, and (iii) the validation of the deployment with a CV application experiment. Although all wireless communication technologies, such as DSRC, LTE, and Wi-Fi are easily accessible by CVs, DSRC is usually preferred for CV safety applications because of its very low communication delay. The performance of heterogeneous wireless technologies, Wi-Fi, DSRC, and LTE was evaluated for the collision avoidance application and traffic data collection and archiving purposes. The field tests found that the heterogeneous wireless network can support multiple applications simultaneously, including safety and mobility applications. For data analytics in real-time data streaming, data aggregation and processing should be done based on on-demand data requests. Brokering systems can support data distribution for real-time data analytics by reducing data redundancy. In addition, real-time data archiving is required at the System Edge level for mobility and safety applications using big data management tools. The experimental results for the identified unique features of the CU-CVT testbed may advance future CV research and serve as reference to agencies for the deployment of CV infrastructure in the real world. The CU-CVT was found to be an invaluable resource for evaluating CV technologies and applications.

Acknowledgments

This material is based on the work supported by the Center for Connected Multimodal Mobility (C^2M^2) (USDOT Tier 1 University Transportation Center) headquartered at Clemson University, Clemson, South Carolina, USA and the National Science Foundation (NSF) under U.S. Ignite Grant #1531127. The authors also wish to acknowledge the editorial assistance of Mr. Godfrey Kimball in the preparation of this draft.

Author Contributions

The authors confirm contribution to the paper as follows: Study conception and design: Mashrur Chowdhury; Data collection: Mizanur Rahman; Anjan Rayamajhi; Sakib Khan; Mhafuzul Islam and Zadid Khan; Analysis and interpretation of results: Mashrur Chowdhury; Mizanur Rahman; Anjan Rayamajhi; Sakib Khan; Mhafuzul Islam, Zadid Khan and James Martin; Draft manuscript preparation: Mashrur Chowdhury; Mizanur Rahman; Anjan Rayamajhi; Sakib Khan; Mhafuzul Islam, Zadid Khan and James Martin. All authors reviewed the results and approved the final version of the manuscript.

References

1. *Connected Vehicle Benefits.* https://www.its.dot.gov/factsheets/pdf/ConnectedVehicleBenefits.pdf. Accessed July 28, 2017.
2. *Department of Transportation, Federal Motor Vehicle Safety Standards; V2V Communications, Notice of Proposed Rulemaking*, January 2017.
3. *Federal Motor Vehicle Safety Standards; V2V Communications.* https://www.federalregister.gov/documents/2017/01/12/2016-31059/federal-motor-vehicle-safety-standards-v2v-communications. Accessed July 28, 2017.
4. Federal Highway Administration. *Vehicle to Infrastructure Deployment Guidance and Products*, 2015. https://www.its.dot.gov/meetings/pdf/V2I_DeploymentGuidanceDraftv9.pdf, Accessed July 28, 2017.
5. Safety Pilot Model Deployment. *Lessons Learned and Recommendations for Future Connected Vehicle Activities.* https://ntl.bts.gov/lib/59000/59300/59361/FHWA-JPO-16-363.pdf. Accessed July 28, 2017.
6. Barbaresso, J., G. Cordahi, D. Garcia, C. Hill, A. Jendzejec, and K. Wright. *USDOT's Intelligent Transportation Systems (ITS) ITS Strategic Plan 2015-2019.* No. FHWA-JPO-14-145, Washington, DC: US Department of Transportation, Intelligent Transportation Systems Joint Program Office, 2014.
7. U.S. Government Accountability Office. *Intelligent Transportation Systems: Vehicle-to-Infrastructure Technologies Expected to Offer Benefits but Deployment Challenges Exist.* GAO-15-775. http://www.gao.gov/products/GAO-15-775. Accessed July 28, 2017.
8. *Affiliated Test Bed Memorandum of Agreement*, version 20140130. https://www.its.dot.gov/presentations/CV_Safety_sept2012/pdf/Day%202%20-%20Affiliated%20Test%20-Bed.pdf. Accessed July 28, 2017.
9. The Ann Arbor Connected Vehicle Test Environment. http://www.aacvte.org/. Accessed July 28, 2017.
10. New York City DOT Pilot. https://www.its.dot.gov/pilots/pilots_nycdot.htm. Accessed July 28, 2017.
11. *Additional information regarding New York City deployment initial Concept of Operation.* http://www.its.dot.gov/pilots/pdf/NYC_ConOpsWebinar.pdf. Accessed July 28, 2017.
12. Wyoming DOT Pilot. https://www.its.dot.gov/pilots/pilots_wydot.htm. Accessed July 28, 2017.
13. Tampa Pilot. https://www.its.dot.gov/pilots/pilots_thea.htm. Accessed July 28, 2017.
14. *California CV TestBed*, California. http://caconnectedvehicletestbed.org/index.php/index.php. Accessed November 9, 2017.
15. Virginia Connected Vehicle Testbed. http://cvi-utc.org/test-bed-development/. Accessed July 28, 2017.
16. Xu, Q., T. Mak T., J. Ko, and R. Sengupta. Vehicle-to-vehicle Safety Messaging in DSRC. *Proc., 1st ACM International Workshop on Vehicular ad hoc Networks*, 2004, pp. 19–28, ACM.
17. *Connected Vehicle Reference Implementation Architecture.* http://www.iteris.com/cvria/html/applications/applications.html. Accessed July 28, 2017.
18. Du, Y., M. Chowdhury, M. Rahman, K. Dey, A. Apon, A. Luckow, and L. B. Ngo. A Distributed Message Delivery Infrastructure for Connected Vehicle Technology Applications. *IEEE Transactions on Intelligent Transportation Systems*, 2017.
19. Rayamajhi, A., M. Rahman, M. Kaur, J. Liu, M. Chowdhury, H. Hu, J. McClendon, K. C. Wang, A. Gosain, and J. Martin. ThinGs In a Fog (TGIF): System Illustration with Connected Vehicles. *Proc., 2017 IEEE 85th Vehicular Technology Conference (VTC Spring)*, Sydney, Australia, 2017.
20. *IEEE Standard for Wireless Access in Vehicular Environments (WAVE) – Networking Services.* IEEE Std 1609.3-2016 (Revision of IEEE Std 1609.3-2010), 2016, pp. 1–160. DOI: 10.1109/IEEESTD.2016.7458115
21. *IEEE Standard for Wireless Access in Vehicular Environments. Multichannel Operation.* https://standards.ieee.org/findstds/standard/1609.4-2016.html. Accessed November 9, 2017.
22. IEEE Approved Draft Standard for Wireless Access in Vehicular Environments–Security Services for Applications and Management Messages - Amendment 1. IEEE P1609.2a/D8, 2017, pp. 1–122.
23. *Security Credential Management System Proof–of–Concept Implementation.* https://www.its.dot.gov/pilots/pdf/SCMS_POC_EE_Requirements.pdf. Accessed November 9, 2017.
24. Ma, Y., Y. Zhou, M. Chowdhury, K.C. Wang, and R.N. Fries. A Framework for Performance Evaluation of Communication Alternatives for Intelligent Transportation Systems. *Journal of Intelligent Transportation System*, Vol. 13, No. 3, 2009, pp. 111–126.
25. Zhou, Y., K. C. Dey, M. Chowdhury, and K. C. Wang. Process for Evaluating the Data Transfer Performance of Wireless Traffic Sensors for Real-time Intelligent Transportation Systems Applications. *IET Intelligent Transport Systems*, Vol. 11, No. 1, 2016, pp. 18–27.

26. Dey, K. C., A. Rayamajhi, M. Chowdhury, P. Bhavsar, and J. Martin. Vehicle-to-vehicle (V2V) and Vehicle-to-infrastructure (V2I) Communication in a Heterogeneous Wireless Network–Performance Evaluation. *Transportation Research Part C: Emerging Technologies*, Vol. 68, 2016, pp. 168–184.
27. Dey, K. C., L. Yan, X. Wang, Y. Wang, H. Shen, M. Chowdhury, L. Yu, C. Qiu, and V. Soundararaj. A Review of Communication, Driver Characteristics, and Controls Aspects of Cooperative Adaptive Cruise Control (CACC). *IEEE Transactions on Intelligent Transportation Systems*, Vol. 17, No. 2, 2016, pp. 491–509.
28. *NoSQL Database*. http://nosql-database.org/. Accessed July 28, 2017.
29. *Database Engine*. https://db-engines.com/en/system/CouchDB%3BGoogle + Cloud + Bigtable%3BMongoDB. Accessed July 28, 2017.
30. MQTT. http://mqtt.org/. Accessed July 28, 2017.
31. Association of State Highway and Transportation Officials. *A Policy on Geometric Design of Highways and Streets*, 6th ed. American Association of State Highway and Transportation Officials, Washington, D.C., 2011.
32. Xu, Q., R. Segupta, D. Jiang, and D. Chrysler. Design and Analysis of Highway Safety Communication Protocol in 5.9 GHz Dedicated Short Range Communication Spectrum. *Proc., the 57th IEEE Semiannual Vehicular Technology Conference*, Vol. 4, 2003, pp. 2451–2455.
33. *Report on Dynamic Speed Harmonization and Queue prediction Algorithm Design*. Final Report – FHWA-JPO-14-168. February 28, 2014. https://ntl.bts.gov/lib/54000/54800/54895/FHWA-JPO-14-168.pdf. Accessed July 28, 2017.

The Standing Committee on Vehicle-Highway Automation (AHB30) peer-reviewed this paper (18-06561).

Any opinions, findings, and conclusions or recommendations expressed in this material are those of the authors and do not necessarily reflect the views of the Center for Connected Multimodal Mobility (C^2M^2) and NSF, and the U.S. Government assumes no liability for the contents or use thereof.

Article

TRR
JOURNAL OF THE TRANSPORTATION RESEARCH BOARD

Transportation Research Record
2018, Vol. 2672(22) 24–35

DOI: 10.1177/0361198118796020
journals.sagepub.com/home/trr
SAGE

Adaptive Cruise Control Operation for Improved Motorway Traffic Flow

Anastasia Spiliopoulou[1], Diamantis Manolis[1], Foteini Vandorou[1], and Markos Papageorgiou[1]

Abstract

This study presents an ACC (adaptive cruise control)–based traffic control strategy which aims to adapt in real time the driving behavior of ACC-equipped vehicles to the prevailing traffic conditions so that motorway traffic flow efficiency is improved. The potential benefits obtained by applying the proposed control concept are demonstrated for different ACC penetration rates by use of validated microscopic simulation applied to a real motorway stretch where recurrent traffic congestion is created under the current manual driving conditions because of an on-ramp bottleneck. The simulation results demonstrate that, even for low penetration rates of ACC vehicles, the proposed control concept improves the average vehicle delay and fuel consumption by reducing the space-time extent of congestion compared with the case of only manually driven or regular ACC vehicles.

The number of cars equipped with adaptive cruise control (ACC) systems has increased in recent years and will further increase in the near future. The ACC systems extend the functionalities of earlier cruise control systems, enabling the drivers to adjust, except for the desired maximum speed, the desired time gap to the leading vehicle when following slower vehicles. Based on the driver's settings and the current measurements, the ACC system determines and applies automatically the appropriate acceleration or deceleration of the ACC-equipped vehicle. At present, ACC systems are mainly designed to increase driving safety and comfort, which may imply conservative values for the ACC system settings (i.e., comparatively large time gaps and low accelerations). For example, although a range of time-gap values is available for the driver to select her desired one, the typical default time-gap value in most equipped vehicles is higher than 2 s. Such conservative parameter values, however, may eventually lead to the degradation of the static and dynamic road capacity compared with conventional manual driving vehicle traffic. The higher the percentage of the ACC vehicles (penetration rate), the more pronounced will be the influence of their driving modus on the overall traffic flow.

In contrast to the vast literature on ACC and CACC (cooperative ACC) systems, which focuses on the design of these systems, there is a comparatively smaller number of works which investigate the impact of ACC and CACC systems on traffic flow. These works aim to capture the impact of ACC on traffic flow under different settings (mainly different time gaps) and penetration rates, using either microscopic simulation (*1–8*) or macroscopic approaches (*9–12*). General conclusions that may be drawn from those studies are that: (i) ACC systems have the potential to improve or deteriorate, depending on their settings, the traffic conditions compared with the case of conventional manually driven vehicles; and (ii) the level of the influence is closely related to the ACC penetration rate. However, the above studies do not systematically examine how the ACC settings should be, actually, specified.

There are some studies (*13*, *14*), which advance in this direction and propose the offline optimization of the (constant) ACC system parameters so as to maximize traffic flow efficiency. They demonstrate that appropriate (optimal) ACC settings may indeed lead to improved traffic performance. Finally, there are a few studies (*15–18*), which go yet a step further and attempt the real-time change of ACC settings, based on the prevailing traffic conditions, to improve the traffic flow efficiency. These studies either consider advisory systems, located in a traffic management center, which provide action advice

[1]Dynamic Systems and Simulation Laboratory, School of Production Engineering and Management, Technical University of Crete, Chania, Greece

Corresponding Author:
Address correspondence to Anastasia Spiliopoulou: natasa@dssl.tuc.gr

(i.e., change of ACC system settings) to vehicles featuring appropriate automation and connectivity capabilities (as in [15, 16]) or consider the autonomous operation of individual ACC vehicles, which, using a vehicle-based detection algorithm, identify the traffic conditions and adapt appropriately their individual ACC settings (as in (17, 18]). Except for real-time traffic data, the corresponding approaches call for network topology information (e.g., on/off-ramp or lane-drop locations) as well for their operation.

In this study, a simple but effective ACC-based control strategy is presented, which aims to adjust in real time the ACC settings of equipped and connected vehicles, in particular the time-gap parameter, based on the prevailing traffic conditions. The main philosophy behind the proposed concept is to: (i) leave the ACC settings untouched at their driver-selected values if traffic flow is clearly under-critical, so as to limit interventions only to traffic situations that call for efficiency increase; and (ii) change the ACC settings gradually as appropriate to improve the flow efficiency when critical traffic states are imminent or present. The proposed control strategy is only dependent on real-time information about the current traffic conditions (no network topology information is required) and is activated only when, where, and to the extent needed. The approach is implemented in a validated real network simulation model and is tested through microscopic simulation. The simulation results demonstrate that the proposed strategy improves the motorway traffic flow efficiency significantly, even for low ACC penetration rates, compared with the case in which the ACC vehicles are driving with their initial, default ACC settings.

Real-Time ACC-Based Traffic Control Concept

According to literature findings, ACC systems have the potential not only to increase driving safety and comfort, but also to smooth traffic flow, decrease fuel consumption, and improve traffic flow efficiency. However, if conservative values are set for their parameters, then the ACC systems may actually lead to a deterioration rather than improvement of traffic flow efficiency. In fact, although drivers of ACC vehicles may be offered a range of 0.8 to 2.2 s (ISO 15622, 2010) to select their desired time gap to the front vehicle, the default time gap is higher than 2 s; moreover, some drivers may be selecting ACC time gaps that are higher than those employed in manual driving. Such time-gap settings would lead to a significant reduction of motorway capacity per lane; this impact can be mitigated if the settings of the ACC vehicles can be updated dynamically in real time through the operation of an ACC-based control strategy.

To illustrate the impact of the time gap on capacity, Figure 2*c* displays a single-lane fundamental diagram for various time gaps, which results in a stationary version of the IDM (intelligent driver model) to be presented in detail later. There is a lane capacity of 1,410 vehicles per hour (vph) for a time gap of 2 s, which is drastically lower than the typical capacity of manually driven vehicle traffic; conversely, a time gap of 0.8 s yields a capacity of 2,830 vph which is drastically higher. Although these values are only rough estimates, they do indicate that, in a mixed context with both manually driven and ACC-driven vehicles, any reduction of the time gap applied by ACC vehicles increases the motorway capacity; and that the adoption of the minimum time gap of 0.8 s would lead to a maximum achievable capacity value for any percentage of ACC vehicles. The proposed control strategy aims to exploit this potential in the presence of connected ACC vehicles by determining in real time the time gaps leading to an increase of the static and dynamic road capacity when, where, and to the extent necessary.

For the presentation of the control strategy, consider a motorway with both manually driven and ACC vehicles. The ACC-vehicle drivers may introduce their desired ACC system settings, desired speed, v_d, and minimum time gap, T_d, but the latter is subject to change if the control strategy recommends or orders a lower time gap. The motorway is considered to be divided into sections, and the traffic management center applies the proposed control strategy at every motorway section $i = 1,2,\ldots$ independently, as illustrated in Figure 1. In particular, at every period (or control interval) t_c, the strategy receives real-time measurements (or estimates) of the exiting flow q_i and mean speed v_i of every section i. This information may be obtained through conventional spot detectors, or may be estimated using vehicle-to-infrastructure (V2I) communication (see, for example, [19]).

In the light of the above discussion, the first goal of the strategy is to attempt the timely increase of the section's static capacity in order to prevent or delay the formation of congestion. To achieve this, the strategy gradually decreases the suggested time gap, when traffic flow is within certain bounds. In particular, the strategy calculates the suggested time gap, $T_{stg,i}$, as a function of the current section flow, $T_i[q_i(k)]$, as shown in Figure 2*a*. This function implies that, as long as the section flow is low (i.e., $q_i \leq Q_1$), the maximum time gap, T_{max}, is suggested, as traffic is not critical. Beyond this lower limit, as the flow increases, the strategy gradually decreases the suggested time-gap value, until for high flow values (i.e., $q_i \geq Q_2$) the strategy suggests the minimum time gap T_{min}. Note that the suggested time-gap value is reduced to the minimum value before the flow reaches the nominal capacity of the section. In this way, the strategy aims to delay, or even prevent, the formation of congestion,

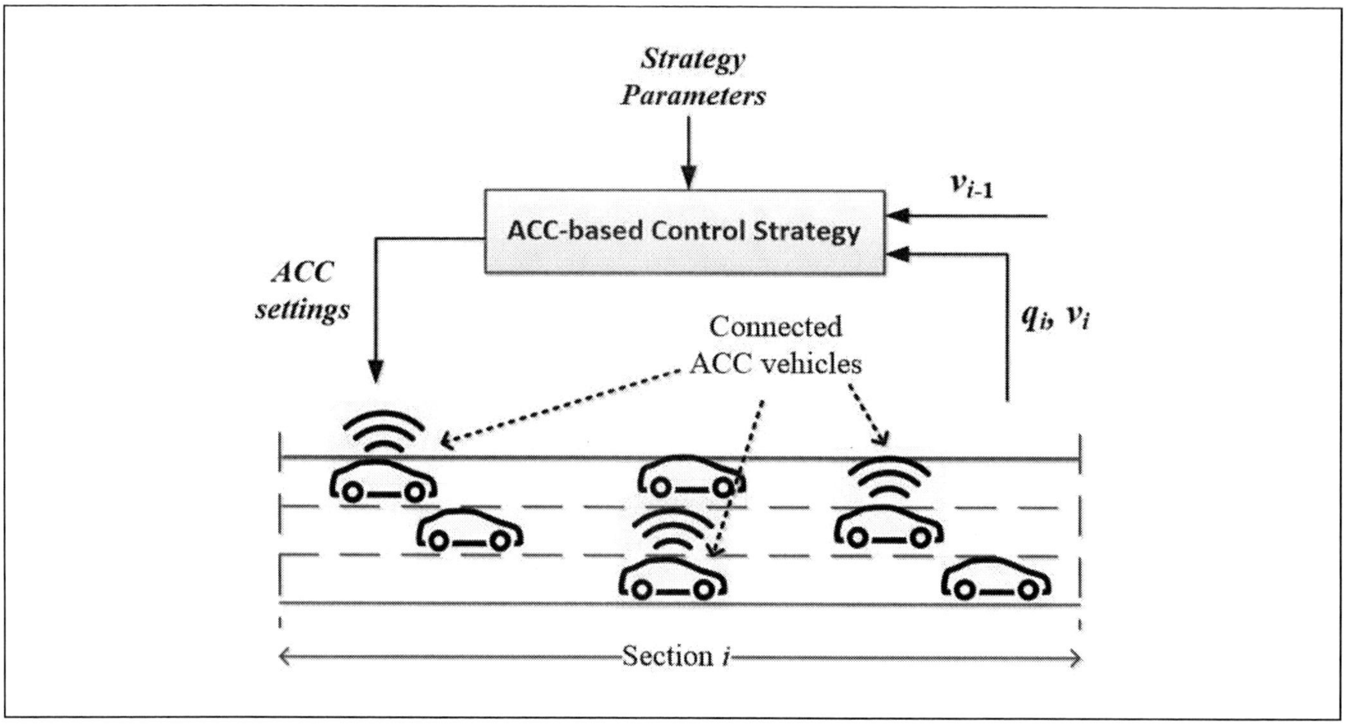

Figure 1. Illustration of the control strategy operation.

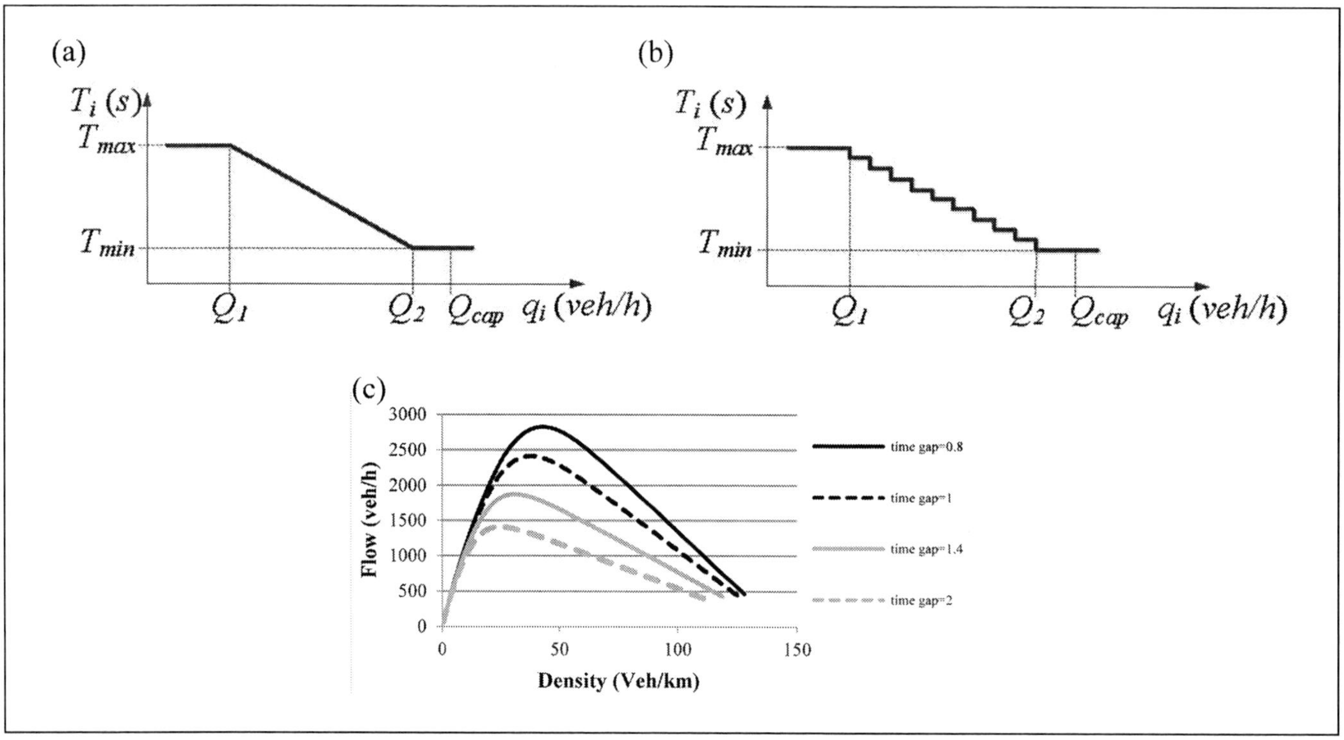

Figure 2. Calculation of the suggested time-gap value using (*a*) a linear function, (*b*) a stepwise function, and (*c*) a single-lane fundamental diagram for various time gaps.

by maximizing in a timely fashion the section's capacity. It should be also noted that the adopted function $T_i[q_i(k)]$ of time gap versus flow must be decreasing, but can have any form deemed appropriate; for example, it may deliver only a (high or low) number of discrete time-gap values, rather than being continuous, as does the stepwise function shown in Figure 2*b*. It is easy to realize that in this case, higher numbers of discrete gap values would lead to more frequent but less abrupt gap changes for the advancing ACC vehicles. If, despite the intended capacity increase, the section becomes congested (e.g., because of even higher arriving demand or because of a shockwave arriving from downstream), then the strategy releases its operation at the congested sections, by suggesting the maximum time gap $T_{\max}$, for safety reasons. Recall that the suggested time gap will be applied only if it is lower than the individual time gap setting. The above control decisions are summarized by the following equation which determines the suggested time gap:

$$T_{\mathrm{stg},i}(k) = \begin{cases} T_i[q_i(k)] & \text{if } v_i(k) > v_{\mathrm{cong}} \\ T_{\max} & \text{else} \end{cases} \qquad (1)$$

where $k = 1,2,\ldots$, is the discrete time index, and v_{cong} indicates the congestion limit. The control strategy decisions are calculated externally, at an infrastructure-based traffic management center (or roadside unit) and are disseminated to the ACC vehicles by a method such as V2I communication.

The second goal of the proposed strategy is the maximization of the discharge flow during congestion at the location of active bottlenecks (dynamic capacity). It is empirically known that the discharge flow at the head of a congested area is lower than capacity, and the second goal is to mitigate this capacity drop. To this end, the strategy first identifies the location of active bottlenecks and then sets the suggested time gap for these sections equal to the minimum time gap $T_{\min}$. More specifically, if two consecutive sections $i-1$ and i, have a speed difference higher than a threshold Δv, and the mean speed $v_i(k)$ of downstream section i is higher than the congestion speed v_{cong}, while the mean speed of the upstream section, $v_{i-1}(k)$, is lower than v_{cong}, this indicates that these two sections are located just upstream and just downstream of the head of a congested area and, therefore, that an active bottleneck has been identified. To achieve the increase of the discharge flow at the congestion head, the strategy suggests the minimum time gap $T_{\min}$ for these two sections. These extended control decisions are implemented via the following relations:

$$\begin{aligned} &\text{IF } v_i(k) > v_{\mathrm{cong}} \text{ AND } v_{i-1}(k) < v_{\mathrm{cong}} \text{ AND } [v_i(k) - v_{i-1}(k)] > \Delta v \\ &\quad T_{\mathrm{stg},i}(k) = T_{\min} \\ &\quad T_{\mathrm{stg},i-1}(k) = T_{\min} \end{aligned} \qquad (2)$$

The employed speed measurements or estimates are sufficiently smoothed before being used in this relation to suppress possible oscillations or false alarms because of moving shock waves.

Equations 1 and 2 determine the suggested time gap $T_{\mathrm{stg},i}$ at each control period. The ACC vehicles receive the suggested time gap, but they apply it only if their individual time-gap setting, $T_{d,j}$, is higher than the time gap calculated by the controller, shown by:

$$T_{\mathrm{applied},j} = \min\{T_{d,j}, T_{\mathrm{stg},i}\} \qquad (3)$$

where $j = 1, 2,\ldots$ is the ACC-vehicle index. Note that the frequency at which the advancing ACC vehicles update their time-gap settings may be higher than at every control interval t_c because of crossing of section boundaries, unless the broadcasting of centrally calculated time gaps is only effectuated strictly every t_c.

The proposed control concept is simple and easy to implement. It considers that the ACC vehicles communicate with a traffic management center which suggests to the drivers or imposes directly appropriate values for the time-gap parameter. It requires only real-time flow and speed measurements (or estimates) from the motorway sections, and is activated only when, where, and to the extent needed. The strategy includes four parameters that should be appropriately specified, which are Q_1, Q_2, v_{cong}, Δv; $T_{\min}$ and $T_{\max}$ are naturally set equal to the bounds of commercial ACC systems. The choice of the Q_1 and Q_2 parameters influences the amount and level of required time-gap changes. The values of v_{cong} and Δv might need some easy fine-tuning for different traffic infrastructures. The sensitivity of results with respect to these parameters was found to be minor in the following investigations.

The employed motorway section length for application of the strategy may depend on the space resolution of the available real-time measurements or estimates. It should be noted that shorter sections tend to reflect more homogeneous traffic conditions and may therefore lead to more pertinent suggested time gaps, for the price of more frequent time-gap changes experienced by advancing ACC vehicles. A similar trade-off applies also to the selection of the strategy's update period. Related simulation investigations indicate low sensitivity of the results near the employed section length and sample time values.

In the investigations presented in the next sections, the potential benefits obtained from the application of the proposed control strategy are examined separately for Equations 1 and 2, to better demonstrate the qualitative features of each ingredient of the proposed concept.

Network Description and Simulation Setup

The simulated network is a part of the A20 motorway in The Netherlands, which connects Rotterdam and Gouda. The motorway stretch considered is about 9.3 km long and includes two on-ramps and two off-ramps. It features three lanes until for 3.6 km, at which point the leftmost

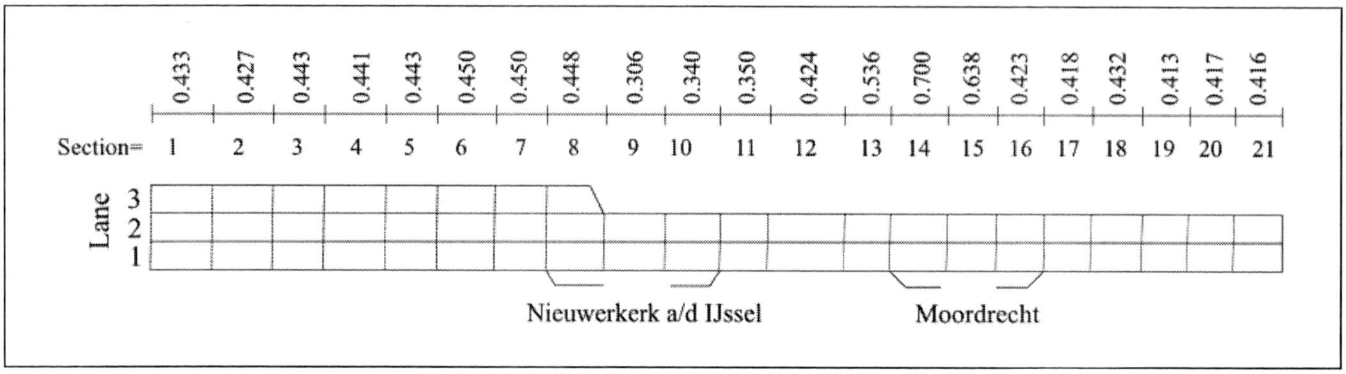

Figure 3. Sketch of motorway A20 in the Netherlands.

lane drops (see Figure 3). Based on the analysis of (*20*), recurrent congestion is created on the examined stretch during the morning peak hours. The trigger of congestion is the high demand on the first on-ramp which, in combination with the mainstream demand, exceeds the motorway capacity. This real motorway stretch is utilized to test and demonstrate the application of the proposed traffic control concept. The microscopic simulator AIMSUN was calibrated and validated using real traffic data so as to reproduce the typical traffic evolution in the examined motorway under manual driving conditions (see [*20, 21*]). Figure 4*a* illustrates the spatiotemporal diagram of the real speed on the examined network, whereas Figure 4*b* illustrates the corresponding diagram obtained through the calibrated simulator. It is observed that the simulator is able to reproduce the formation of congestion at the first on-ramp for the correct time period and extent. Note that this study focuses on the bottleneck caused by the on-ramp merge, so the second wave of congestion that enters the examined motorway stretch from the downstream boundary (see Figure 4*a*) is out of the area of interest for this work and is therefore not replicated in the simulated model which assumes free-flow traffic conditions downstream of the motorway stretch at all times. Moreover, Figure 4*c* presents the time trajectory of the outflow from the congested area, which is downstream of the first on-ramp; it demonstrates that, during the congestion period, the outflow, in both the real and the simulated network, is reduced below the motorway capacity (capacity drop). The calibration of the simulation model was carried out considering only manually driven vehicles. This corresponds to the base (no–traffic control) case of the presented investigations. Then, the proposed control concept was applied examining different cases with regard to the proportion of vehicles equipped with ACC technology.

Both manual and ACC vehicles are moving according to the IDM (*22*), which is a simple, yet realistic, car-following model. The IDM calculates the vehicle acceleration $\dot{v}$ based on:

$$\dot{v}(s, v, \Delta v) = a\left[1 - \left(\frac{v}{v_0}\right)^4 - \left(\frac{s^*(v, \Delta v)}{s}\right)^2\right] \quad (4)$$

where

a is the maximum acceleration;
v_0 is the desired speed;
s is the current distance to the preceding vehicle; and
$s^*(v,\Delta v)$ is the desired distance which is calculated by:

$$s^*(v, \Delta v) = s_0 + vT + \frac{v\Delta v}{2\sqrt{ab}} \quad (5)$$

where

s_0 is the minimum distance;
T is the minimum time gap;
and b is the comfortable deceleration.

Note that all IDM parameters, namely, v_0, T, s_0, a, and b, take positive values. Within the AIMSUN simulator, these parameters are defined by a mean, maximum and minimum value; individual vehicles are sampled from a truncated normal distribution. In the manual driving case, the mentioned parameter calibration yielded the range $T \in [0.68, 1.67]$ for the time-gap parameter with mean value 1.15 s and standard deviation 0.26 s; see (*20, 21*) for other details.

Real ACC algorithms implemented in equipped vehicles are complex, and their details are different across car manufacturers and unknown. As a substitute for microscopic traffic simulation studies, simple dynamic formulas have been utilized in various works, including the IDM in some of them. How to simplify reliably the ACC-vehicle dynamics in a microscopic traffic simulation environment seems to be an open issue. The main interest of the present investigations is the impact of the time gap; the IDM equation is therefore also employed for ACC vehicles, the only difference between the driving characteristics of manual and ACC vehicles being the respective utilized distribution bounds for the time-gap parameter T. In particular, for the ACC vehicles,

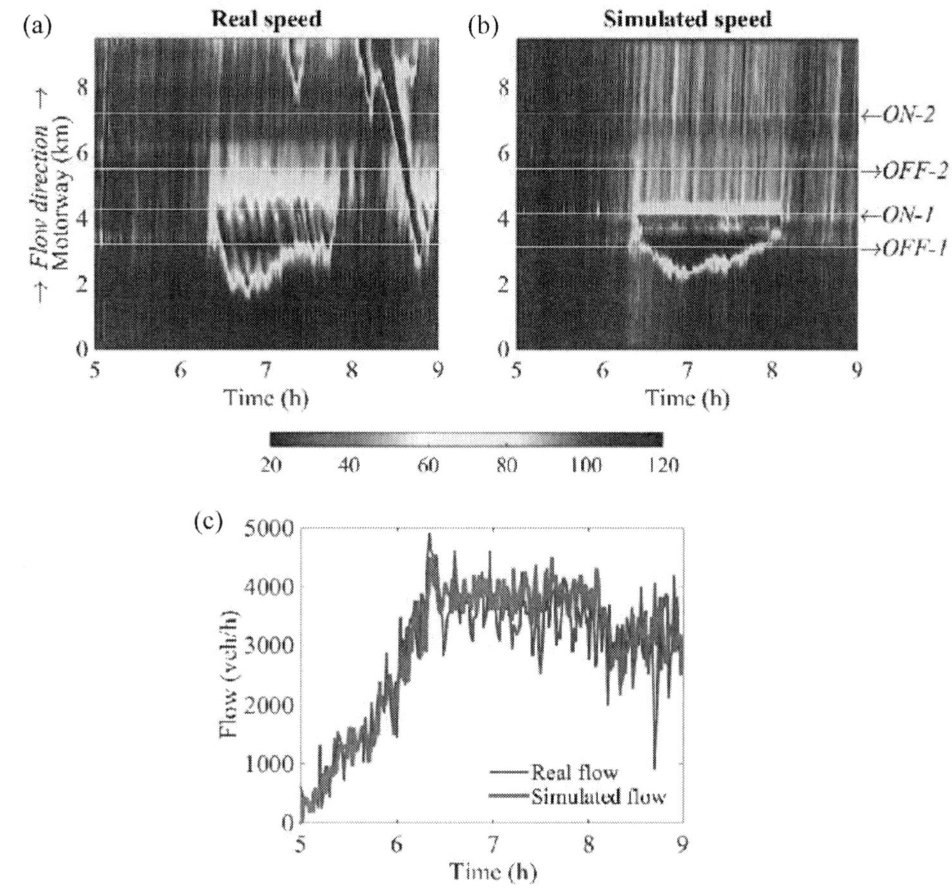

Figure 4. (*a*) Spatiotemporal diagrams of real speed; (*b*) spatiotemporal diagrams of simulated speed; (*c*) real and simulated outflow from the merge area.

$T \in [0.8, 2.2]$ s (again with a truncated normal distribution) with mean value 1.4 s and standard deviation 0.3 s. Thus, the range of time-gap values for manual vehicles was derived from the calibration procedure and corresponds to typical values for manual driving, whereas the mentioned range of values for ACC vehicles delivers a mean value that is comparable with the one found empirically to be employed by the users (*23*). The utilized AIMSUN simulation step was set to 0.4 s, and the control period was set to 30 s. IDM was chosen for both manual and ACC vehicles, to ensure that the observed effects are caused by the time-gap policy and not by a different ACC dynamic vehicle model.

Simulation Investigations

In the following sections, the simulation results of the no–traffic control case are presented first, followed by the results of applying the first part of the proposed control strategy (Equation 1) and its complete version (Equations 1 and 2), considering various penetration rates. Ten replications were carried out for each examined scenario, and the average values of the performance criteria are compared. The reported fuel consumption and delay criteria are delivered by AIMSUN.

No–Traffic Control Case

In the case without traffic control, 100% manually driven vehicles are considered. Table 1 includes the average vehicle delay (AVD) in seconds per vehicle per kilometer (s/veh/km) and the total fuel consumption (TFC) in liters (L), which are the mean values over 10 replications and are equal to 19.1 s/veh/km and 6,551 respectively. Figure 5 illustrates the traffic situation on the motorway for 0% penetration rate, for a particular replication with AVD value closest to the mean value of the 10 replications. As mentioned before, it is observed that congestion is formed between 6:30 a.m. and 8:00 a.m. at the first on-ramp, which propagates upstream for about 2 km.

Table 1. Traffic Control and No–Traffic Control Cases: AVD, TFC, and Comfort Criteria Considering Various Penetration Rates

Performance indices		Penetration rate					
		0%	5%	10%	20%	30%	60%
AVD (s/veh/km)	No–traffic control	19.1	20.3	23.7	27.1	32.7	45.1
	Control case 1	na	18.0	19.3	19.4	14.3	13.7
	Difference (%)	na	−11.4	−18.6	−28.3	−56.3	−69.6
	Complete control case	na	15.8	13.5	11.3	7.9	6.5
	Difference (%)	na	−22.5	−42.8	−58.3	−75.8	−85.6
TFC (L)	No–traffic control	6,551	6,577	6,654	6,717	6,812	6,731
	Control case 1	na	6,463	6,535	6,524	6,229	6,060
	Difference (%)	na	−1.7	−1.8	−2.9	−8.6	−9.9
	Complete control case	na	6,377	6,307	6,182	5,938	5,829
	Difference (%)	na	−3.0	−5.2	−8.0	−12.8	−14.6
% time of having different time gap than desired	Control case 1	na	36.3%	35.5%	34.7%	35.7%	33.1%
	Complete control case	na	40.6%	40.8%	40.4%	38.7%	34.5%

Note: na = not applicable.

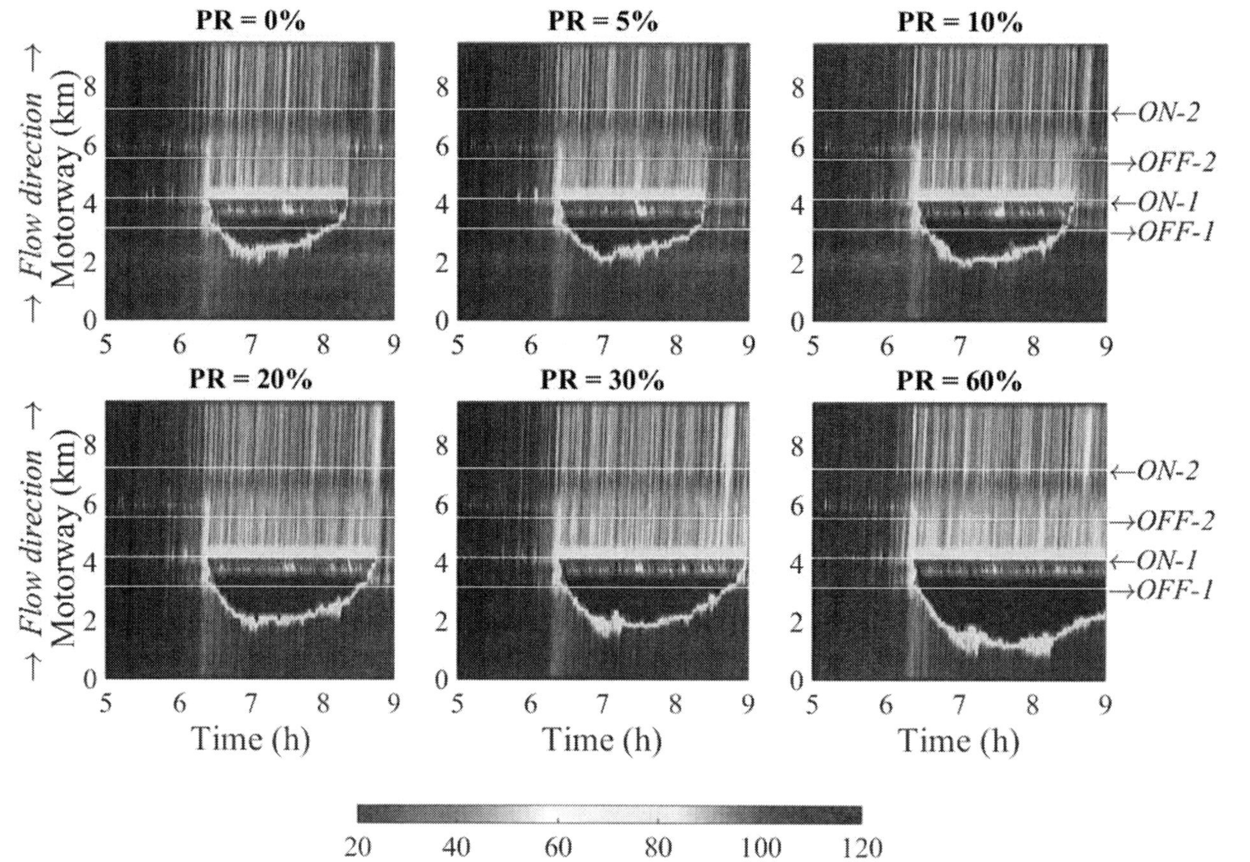

Figure 5. No–traffic control case: spatiotemporal diagrams of speed considering various penetration rates (PR).

As discussed earlier, the utilized range of values for the ACC time-gap parameter is [0.8, 2.2] s, the mean value of which is higher than the calibrated mean value of manually driven vehicles. This implies that the introduction of ACC vehicles may reduce the traffic flow efficiency. To investigate this issue, a series of simulation

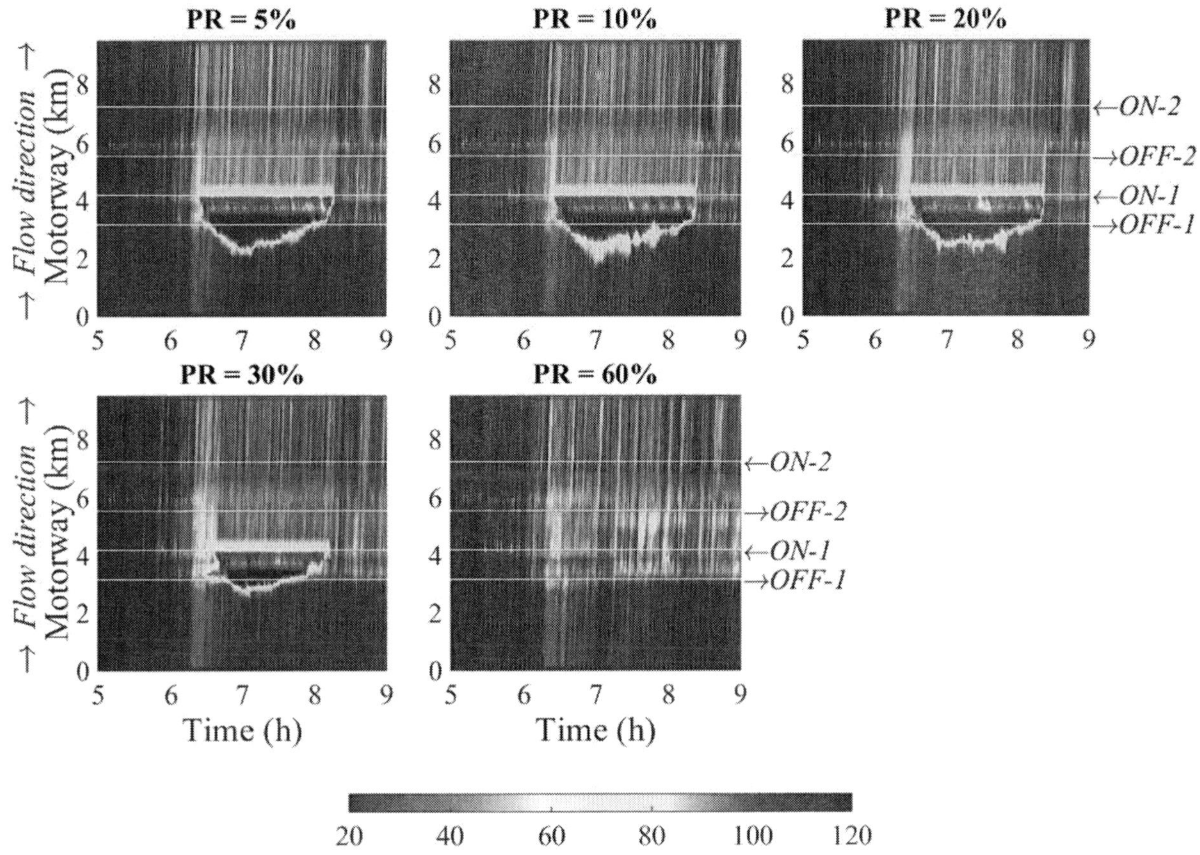

Figure 6. Traffic control case 1: spatiotemporal diagrams of speed considering various penetration rates (PR).

runs were carried out considering various respective penetration rates of ACC vehicles. Table 1 presents, for each investigated penetration rate, the mean AVD and TFC values over 10 replications. It may be seen that, the higher the penetration rate, the bigger the mean AVD and TFC values, which reflect the deterioration of the network traffic conditions, as illustrated also in Figure 5. These findings reject the perception that the introduction of ACC vehicles will automatically improve the driving conditions, as it demonstrates that the "uncontrolled" behavior of ACC vehicles may actually contribute to further degradation of the road infrastructure.

Traffic Control Case 1

The proposed traffic control concept was applied to the investigated network. In the following, the application results of the first part (Equation 1) of the control strategy, which aims at increasing the static road capacity, are presented; the simulation results of the complete (Equations 1 and 2) control strategy are presented in the next section.

The control strategy is activated every $t_c = 30$ s, receiving real-time measurements of flow and speed for every section of the motorway (see Figure 1), whereby section lengths are around 500 m. The parameters involved in the strategy equations are set as follows: $v_{\text{cong}} = 60$ km/h, $T_{\min} = 0.8$ s, $T_{\max} = 2.2$ s, $Q_1 = 1{,}250$ vph per lane, $Q_2 = 1{,}700$ vph per lane. The strategy calculates the suggested time gap T_{stg} based on Equation 1, using discrete time-gap values from the range [0.8, 2.2] *s* with increments of 0.1 *s* (see Figure 2*b*). The ACC vehicles receive, every simulation step, the strategy's decisions and update their individual time-gap setting according to Equation 3.

Table 1 presents the control simulation results in terms of mean AVD and TFC for various penetration rates, along with the average percentage of time that ACC vehicles drive with lower than desired time gap. It is observed that, the higher the penetration rate, the higher the improvement in mean AVD and TFC. Note in particular that even for penetration rate as low as 5%, a considerable improvement of the traffic conditions is achieved, corresponding to 11.4% lower AVD and 1.7%

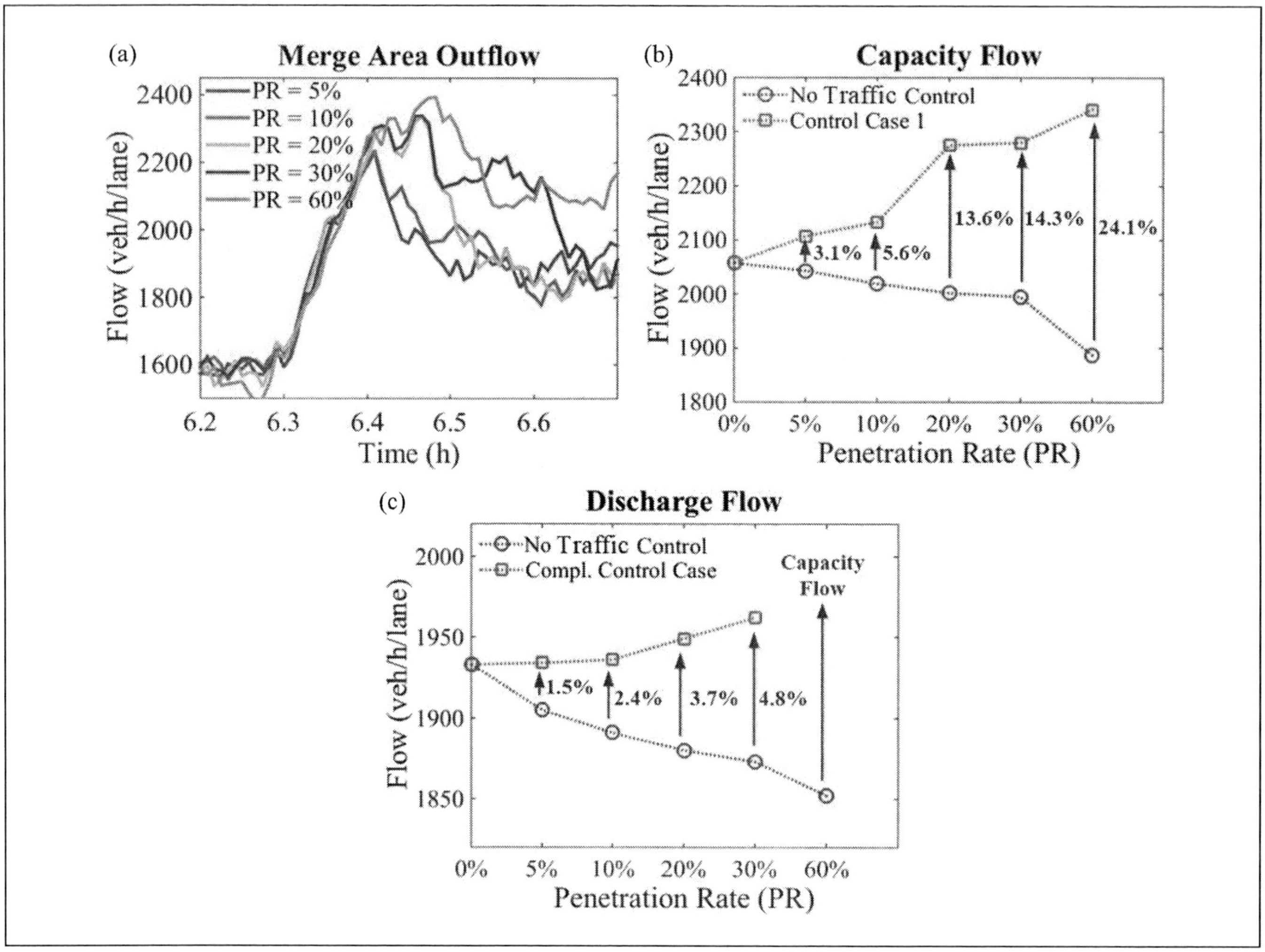

Figure 7. (*a*) Time trajectory of the outflow from the merge (congested) area; (*b*) capacity flows for the no–traffic control and traffic control cases 1; (c) discharge rates for the no–traffic control and complete traffic control cases, for various penetration rates.

less TFC, compared with the corresponding no-control case of 5% ACC vehicles. On the other hand, for a penetration rate equal to 60%, the control strategy manages to resolve congestion completely for the majority of replications. Figure 6 displays the spatiotemporal diagrams of speed for various penetration rates. Each diagram corresponds to a particular replication with AVD value closest to the mean AVD value of the respective 10 replications. Again, it may be seen that, the higher the penetration rate, the bigger the improvement of the traffic conditions on the motorway stretch.

As mentioned before, the first goal of the proposed control strategy is the increase of the static road capacity via Equation 1. Figure 7*a* presents the trajectories of the outflow from the merge area around the onset of congestion for various penetration rates. Again, the presented results correspond to the replications with AVD values closest to the mean value of the respective10 replications. Figure 7*a* illustrates that, for higher penetration rates, the onset of congestion is increasingly delayed, as in fact the capacity of the merge area is seen to increase. Moreover, Figure 7*b* illustrates the capacity flows reached at the no–traffic control and control cases for each corresponding penetration rate. Each point on the graph corresponds to a 5-min average of flow before breakdown. It is observed that, in the no–traffic control case, the higher the penetration rate of ACC vehicles, the lower the capacity flow reached because of the bigger time gaps kept between the vehicles. On the other hand, the application of the proposed control strategy manages to increase the static road capacity, whereby, the higher the penetration rate, the bigger the achieved improvement.

Complete Traffic Control Case

The second goal of the proposed strategy is the maximization of discharge flow rate during congestion at the

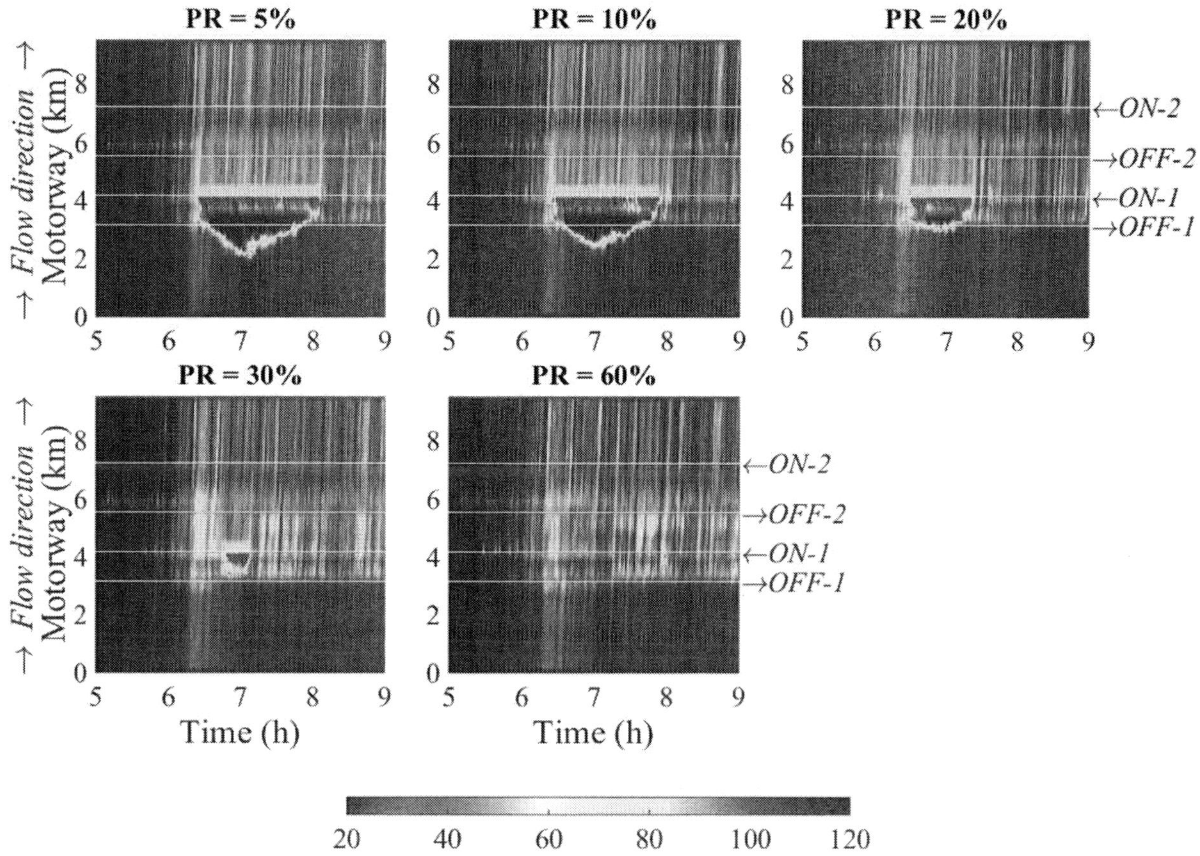

Figure 8. Complete traffic control case: spatiotemporal diagrams of speed considering various penetration rates.

location of active bottlenecks. To this end, Equation 2 is employed in addition to Equation 1. In particular, the complete strategy identifies the location of active bottlenecks and sets the suggested time gap for those sections equal to the minimum time gap $T_{\min}$. The additional parameter Δv was set equal to 10 km/h. It should be noted that, with these parameter settings, the conditions employed in Equation 2 identify the active bottleneck promptly and robustly.

Table 1 presents the simulation results of the complete strategy in terms of mean AVD and TFC for various penetration rates, again with the average percentage of time driving with lower than desired ACC time gap. It is observed that the higher the penetration rate, the higher the improvement in mean AVD and TFC. Compared with the partial control concept (Equation 1 only), the complete strategy achieves almost double improvement, especially for penetration rates up to 20%, and for penetration rate equal to 60%, the complete strategy succeeds in completely resolving congestion in all 10 replications. Note that the calculated improvements would be even higher for a tighter space-time window around the congestion. Figure 8 presents the spatiotemporal diagrams of speed for various penetration rates. Again, it may be seen that the improvement of traffic conditions is bigger compared with the partial control strategy (compare with Figure 6), especially for low penetration rates.

Finally, Figure 7c illustrates the average discharge flow during congestion, in the no–traffic control and complete control cases, for each corresponding penetration rate. Each point on the graph corresponds to a 20-min average of discharge flow during congestion. It is observed that in the no–traffic control case, the higher the penetration rate of ACC vehicles, the lower the discharge flows reached during congestion, whereas with application of control, the corresponding discharge flows are increased; and, again, the higher the penetration rate, the bigger the achieved improvement—for penetration rate equal to 60%, the congestion on the motorway stretch is fully avoided.

Summarizing the obtained evaluation results, it is worth highlighting the specific and distinct role of each part of the proposed control strategy:

- Equation 1 is used to increase the static capacity at bottleneck locations, and the level of increase depends on the penetration rate of ACC vehicles. If the achievable capacity increase is sufficient to accommodate the arriving demand, then the activation of the bottleneck may be avoided altogether. For the present case study, this actually happens only for most replications of the case with a penetration rate of 60%. In all other cases, in which the arriving demand is higher than the increased capacity, the congestion forming can merely be delayed, leading to a reduction of the congestion extent and corresponding traffic improvements. After the onset of congestion, however, Equation 1 is essentially inactive.
- The second part of the proposed strategy, Equation 2, only makes a contribution if congestion actually forms; else, the conditions of Equation 2 will not be satisfied, and this part of the control strategy remains inactive. However, when this strategy part is actually employed (because congestion has formed), then its contribution is significant, because the increase of the discharge flow leads to accordingly faster resolution of the congestion with corresponding benefits for traffic flow efficiency.

Conclusions

This paper presents an innovative ACC-based traffic control strategy, which aims to dynamically adapt the driving behavior of ACC-equipped vehicles so that the motorway traffic flow efficiency is improved. The proposed strategy was tested through microscopic simulation using a real traffic motorway stretch where congestion is created by an on-ramp bottleneck. The simulation results showed that, the higher the penetration rate, the bigger the improvement of the prevailing traffic conditions, for both the AVD and the TFC.

IDM has been used for both manual and ACC vehicles to ensure that the mentioned results reflect the time-gap effects on the traffic. Microscopic simulation is useful in this sort of investigation, but, because of several uncertainties about the dynamics of ACC vehicles, the produced simulation-based traffic-level comparisons and improvements might turn out to be quantitatively different if the proposed concept were to be applied in real conditions. However, the obtained results are useful to demonstrate and evaluate the impact tendency and pertinence of the concept.

In this research work, only the dynamic adaptation of the time-gap settings was considered, whereas ongoing research includes the dynamic adaptation of other ACC system settings as well.

Acknowledgments

The research leading to these results has received funding from the European Research Council under the European Union's Seventh Framework Programme (FP/2007-2013)/ERC Grant Agreement n. [321132], project TRAMAN21. The authors would like to thank Prof. Bart van Arem and his group for their support in providing information and data related to the network used in the application example.

Author Contributions

The authors confirm contribution to the paper as follows: study conception and design: MP, AS; data collection: FV; analysis and interpretation of results: AS, DM, FV; draft manuscript preparation: AS, DM. All authors reviewed the results and approved the final version of the manuscript.

References

1. Vanderwerf, J., S. Shladover, N. Kourjanskaia, M. Miller, and H. Krishnan. Modeling Effects of Driver Control Assistance Systems on Traffic. *Transportation Research Record: Journal of the Transportation Research Board*, 2001. 1748: 167–174.
2. Bose, A., and P. Ioannou. Analysis of Traffic Flow with Mixed Manual and Semi Automated Vehicles. *IEEE Transactions on Intelligent Transportation Systems*, Vol. 4, No. 4, 2003, pp. 173–188.
3. Ioannou, P., and M. Stefanovic. Evaluation of the ACC Vehicles in Mixed Traffic: Lane Change Effects and Sensitivity Analysis. *IEEE Transactions on Intelligent Transportation Systems*, Vol. 6, No. 1, 2005, pp. 79–89.
4. Van Arem, B., C. J. G. Van Driel, and R. Visser. The Impact of Cooperative Adaptive Cruise Control on Traffic Flow Characteristics. *IEEE Transactions on Intelligent Transportation Systems*, Vol. 7, No. 4, 2006, pp. 429–436.
5. Schakel, W. J., B. Van Arem, and B. D. Netten. Effects of Cooperative Adaptive Cruise Control on Traffic Flow Stability. *Proc., 13th International IEEE Conference on Intelligent Transportation Systems*, 2010, pp. 759–764.
6. Shladover, S. E., D. Su, and X.-Y. Lu. Impacts of Cooperative Adaptive Cruise Control on Freeway Traffic Flow. *Transportation Research Record: Journal of the Transportation Research Board*, 2012. 2324: 63–70.
7. Ntousakis, I., I. K. Nikolos, and M. Papageorgiou. Procedia Social and Sciences International *Symposium of Transport Simulation* 2014 on Microscopic Modelling of Adaptive Cruise Control Systems. International Symposium of Transport Simulation, 2014, pp. 1–19.
8. Bayar, B., S. A. Sajadi-Alamdari, F. Viti, and H. Voos. Impact of Different Spacing Policies for Adaptive Cruise Control on Traffic and Energy Consumption of Electric Vehicles. *Proc., 24th Mediterranean Conference on Control and Automation*, 2016, pp. 1349–1354.
9. Ngoduy, D. Application of Gas-Kinetic Theory to Modelling Mixed Traffic of Manual and ACC Vehicles. *Transportmetrica*, Vol. 8, No. 1, 2012, pp. 43–60.
10. Ngoduy, D. Instability of Cooperative Adaptive Cruise Control Traffic Flow: A Macroscopic Approach.

Communications in Nonlinear Science and Numerical Simulation, Vol. 18, No. 10, 2013, pp. 2838–2851.
11. Nikolos, I. K., A. I. Delis, and M. Papageorgiou. Macroscopic Modelling and Simulation of ACC and CACC Traffic. *Proc., 18th IEEE International Conference on Intelligent Transportation Systems*, 2015, pp. 2129–2134.
12. Delis, A. I., I. K. Nikolos, and M. Papageorgiou. Simulation of the Penetration Rate Effects of ACC and CACC on Macroscopic Traffic Dynamics. *Proc., 19th IEEE International Conference on Intelligent Transportation Systems*, 2016, pp. 336–341.
13. Mba, C. U., and C. Novara. Evaluation and Optimization of Adaptive Cruise Control Policies via Numerical Simulations. *2nd International Conference on Vehicle Technology and Intelligent Transport Systems VEHITS*, 2016, pp. 13–19.
14. Liu, H., H. Wei, T. Zuo, Z. Li, and Y. J. Yang. Fine-Tuning ADAS Algorithm Parameters for Optimizing Traffic Safety and Mobility in Connected Vehicle Environment. *Transportation Research Part C: Emerging Technologies*, Vol. 76, 2017, pp. 132–149.
15. Schakel, W., G. Klunder, B. van Arem, E. Harmsen, and M. Hagenzieker. Reducing Travel Delay by In-Car Advice on Speed, Headway and Lane Use Based on Downstream Traffic Flow Conditions: A Simulation Study. *Proc., 15th Meeting of the Euro Working Group on Transportation EWGT*, 2012.
16. Schakel, W. J., and B. van Arem. Improving Traffic Flow Efficiency by In-Car Advice on Lane, Speed, and Headway. *IEEE Transactions on Intelligent Transportation Systems*, Vol. 15, No. 4, 2014, pp. 1597–1606.
17. Kesting, A., M. Treiber, M. Schönhof, and D. Helbing. Adaptive Cruise Control Design for Active Congestion Avoidance. *Transportation Research Part C: Emerging Technologies*, Vol. 16, No. 6, 2008, pp. 668–683.
18. Kesting, A., M. Treiber, and D. Helbing. Enhanced Intelligent Driver Model to Access the Impact of Driving Strategies on Traffic Capacity. *Philosophical Transactions of the Royal Society Series A: Mathematical, Physical and Engineering Sciences*, Vol. 368, No. 1928, 2010, pp. 4585–4605.
19. Bekiaris-Liberis, N., C. Roncoli, and M. Papageorgiou. Highway Traffic State Estimation with Mixed Connected and Conventional Vehicles. *IEEE Transactions on Intelligent Transportation Systems*, Vol. 17, 2016, pp. 3484–3497.
20. Perraki, G. *Evaluation of a Model Predictive Control Strategy on a Calibrated Multilane Microscopic Model.* MSc thesis. Technical University of Crete, 2016.
21. Perraki, G., C. Roncoli, I. Papamichail, and M. Papageorgiou. Evaluation of an MPC Strategy for Motorway Traffic Comprising Connected and Automated Vehicles. *20th IEEE 20th International Conference on Intelligent Transportation Systems*, 2017. submitted.
22. Treiber, M., A. Hennecke, and D. Helbing. Congested Traffic States in Empirical Observations and Microscopic Simulations. *Physical Review E*, Vol. 62, No. 2, 2000, pp. 1805–1824.
23. Nowakowski, C., S. E. Shladover, C. Delphine, F. Bu, J. O'Connell, J. Spring, S. Dickey, and D. Nelson. *Cooperative Adaptive Cruise Control: Testing Drivers' Choice of Following Distances.* California PATH Research Report. Berkeley, CA, 2011.

The Standing Committee on Vehicle-Highway Automation (AHB30) peer-reviewed this paper (18-00930).

Article

TRR
JOURNAL OF THE TRANSPORTATION RESEARCH BOARD

Transportation Research Record
2018, Vol. 2672(22) 36–46

DOI: 10.1177/0361198118793001
journals.sagepub.com/home/trr
SAGE

Hardware-in-the-Loop Testing of Connected and Automated Vehicle Applications: A Use Case for Queue-Aware Signalized Intersection Approach and Departure

Jiaqi Ma[1], Fang Zhou[2], Zhitong Huang[3], Christopher L. Melson[4], Rachel James[4], and Xiaoxiao Zhang[5]

Abstract

Most existing studies on connected and automated vehicle (CAV) applications apply simulation to evaluate system effectiveness. Model accuracy, limited data for calibration, and simulation assumptions limit the validity of evaluation results. One alternative approach is to use emerging hardware-in-the-loop (HIL) testing methods. HIL test environments enable physical test vehicles to interact with virtual vehicles from traffic simulation models, providing an evaluation environment that can replicate deployment conditions at early stages of CAV technology implementation without incurring excessive costs related to large field tests. In this study, a HIL testing system for vehicle-to-infrastructure (V2I) CAV applications is developed. The involved software and hardware includes a physical CAV controlled in real time, a traffic signal controller, communication devices, and a traffic simulator (VISSIM). Such HIL systems increase validity by considering the physical vehicle's trajectories—which are constrained by real-world factors such as GPS accuracy, communication delay, and vehicle dynamics—in a simulated traffic environment. The developed HIL system is applied to test a representative early deployment CAV application: queue-aware signalized intersection approach and departure (Q-SIAD). The Q-SIAD algorithm generates recommended speed profiles based on the vehicle's status, signal phase and timing (SPaT), downstream queue length, and system constraints and parameters (e.g., maximum acceleration and deceleration). The algorithm also considers the status of other vehicles in designing the speed profiles. The experiment successfully demonstrated this functionality with one test CAV driving through one intersection controlled by a fixed-timing traffic signal under various simulated traffic conditions.

Connected and automated vehicle (CAV) technologies offer potentially transformative traffic impacts, including mobility, safety, and environmental benefits. Numerous modeling studies have been conducted to evaluate these impacts. However, model accuracy and simulation assumptions limit the validity of evaluation results; additionally, the lack of field data exacerbates this problem of inaccuracy, leading to improper model calibration. These factors have resulted in the wide range of differences in effectiveness among studies examining the same CAV applications available in the literature.

Some studies use field experiments to circumvent the aforementioned limitations to investigate whether these new technologies can function as expected in practice and to collect data on system performance under various conditions (e.g., mixed traffic, system delay, or inaccurate input data). However, large-scale field operational tests—such as the 100-car autonomous-vehicle tests planned in Sweden and China (*1, 2*) and the three pilot deployments of connected vehicles in the United States (*3*)—are extremely expensive and still cannot support full-scale assessments of the effect that CAVs will have on overall traffic performance. There are also many confounding

[1]University of Cincinnati, Cincinnati, OH
[2]StoneRiver Inc., CT
[3]Leidos Inc., Reston, VA
[4]Federal Highway Administration, Washington, D.C.
[5]University of Virginia, Charlottesville, VA

Corresponding Author:
Address correspondence to Zhitong Huang: zhitong.huang@leidos.com

factors in a real-world field experiment and possible testing scenarios are limited.

One approach to overcome these challenges is to use emerging hardware-in-the-loop (HIL) tools that allow physical test vehicles to interact with virtual vehicles from traffic simulation models. This provides an evaluation environment that can replicate actual conditions at early stages of CAV deployment without incurring excessive costs. Integrating physical CAVs as hardware in an HIL system is a necessary next step for CAV application evaluation and brings many benefits.

This study develops a HIL testing system for CAV applications in a vehicle-to-infrastructure (V2I) environment. The system includes a physical CAV, dedicated short-range communication (DSRC) equipment (i.e., onboard units (OBUs) and roadside units (RSUs)), a physical traffic controller, a traffic simulator, and a field test track. The developed HIL system can be applied for the following purposes:

- Test and verify CAV applications prior to field experiments;
- Quantify potential benefits of CAV applications in different operating environments (e.g., CAV market penetration rate);
- Conduct testing of different hardware and software components.
- A V2I application, queue-aware signalized intersection approach and departure (Q-SIAD), is considered as a use case for the developed HIL testing architecture. Thus, the contributions of this paper are as follows:
- Development of an HIL testing architecture for V2I-based CAV applications;
- Implementation of the proposed HIL architecture at a selected CAV testbed;
- Testing of a selected CAV application, Q-SIAD, to observe system performance and obtain an understanding of Q-SIAD effectiveness through the developed HIL.

Following this introduction, this paper reviews and analyzes previous studies related to applicable HIL systems. After a brief introduction of the Q-SIAD algorithm, the next section is dedicated to the introduction of the HIL system architecture and testing facility. The following section introduces experiment design and testing results. The last section summarizes findings and suggests future research directions.

Literature Review

The advantages of HIL systems include low cost, shorter experimental life cycle, safe testing environment, and high fidelity (*4*). Three major types of hardware are used in HIL systems for transportation research and are described below.

Traffic Signal Controller

Microscopic simulation is the primary method by which researchers evaluate the effectiveness of traffic control strategies. Assumptions on traffic controllers may not be realistic, however, as vendors are not willing to share the exact algorithms used in their products. Therefore, it is not feasible to duplicate their control algorithms in microscopic simulations, and it is preferable to integrate a physical traffic controller into the simulation. The most basic HIL architecture is comprised of three components: a traffic simulator, a physical signal controller, and a controller interface device. This typical platform is widely used by researchers to test different traffic control logic (e.g., (*5–7*)).

Driving Simulator

To understand human driving behavior, its causes, and its effects on surrounding traffic better, driving simulators are used with simulation software (also called "human-in-the-loop" systems). By observing drivers' reactions and collecting data through experiments, researchers can better understand the effects of human behavior under certain traffic conditions. For example, previous studies have investigated the relationship between distraction from in-car activities and the crash rate using human-in-the-loop simulators (*8, 9*).

Automotive Components

HIL simulation has been widely used to develop and calibrate automotive components. Although parameters related to vehicle dynamics can be set by users in a microscopic traffic simulator, it is preferable for physical automotive components to be integrated in the simulation. Physical automotive components represent actual vehicle dynamics. One representative HIL tool for this purpose was developed jointly by the University of Minnesota and University of Michigan (*10*). A laboratory powertrain research platform, which consists of a physical engine, an engine-loading device, and a virtual powertrain model to represent a vehicle were connected remotely to a microscopic traffic simulator. Based on this system, actual fuel and emissions measurements, which otherwise could not be calculated precisely by fuel and emission maps in simulation, were achieved by the HIL tool.

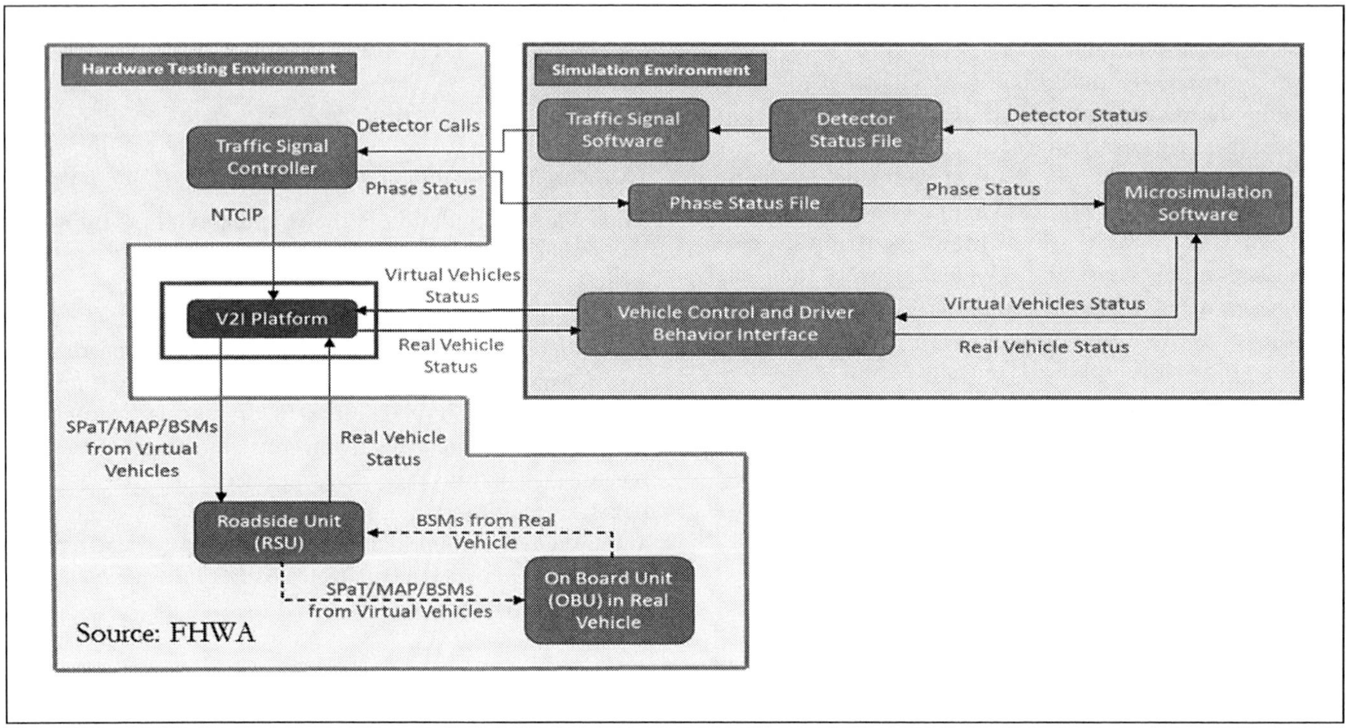

Figure 1. Framework and data flows of the developed hardware-in-the-loop system.

Integrated Testing Platforms

Integrated testing platforms that contain multiple pieces of hardware and software have also attracted researchers' attention. These platforms offer greater convenience by allowing researchers to select the required hardware and software as needed. The State University of New York at Buffalo developed the integrated traffic–driving–networking simulation, which includes three elements: driving simulator, network simulation, and traffic simulation (*11*). Texas Transportation Institute has developed one of the most comprehensive HIL platforms for CAV applications. The system consists of four components: traffic simulation environment, communication simulation environment, applications testing environment, and DSRC hardware testing environment (*12*). These integrated systems, however, focus only on testing technologies related to CVs.

At the time of writing, no literature was found regarding comprehensive HIL systems that can evaluate CAV applications—with a focus on automation—and conduct testing under different traffic and operating conditions. The development of the HIL testing system for CAV applications in a V2I environment described in the remainder of this paper seeks to fill this void.

Hardware-in-the-Loop Testing System

This section describes the new HIL system developed in this work. The system was developed with a V2I focus, but can be modified to evaluate CAV applications based on vehicle-to-vehicle (V2V) technologies.

System Architecture

Figure 1 shows the system framework and data flow between different components of the proposed CAV HIL testing system. The blue box indicates the hardware modules of the HIL system. The software components are identified in the orange box. Note that the interfaces in the orange box refer to software that converts data format for input into next modules. Figure 1 uses dashed lines and solid lines to indicate DSRC and ethernet communication, respectively. A traffic simulator, VISSIM, is used in this system to generate virtual background traffic. VISSIM sends all virtual vehicles' status data and detector data to the HIL interface, which then passes these data to the V2I hub. The V2I hub, developed by FHWA, is an open-source DSRC data hub that encodes and decodes DSRC data sent from vehicles (e.g., basic safety messages (BSMs)) and the infrastructure components (i.e., SPaT messages). The V2I hub is a message handler that acts as a translator and data aggregator or disseminator for the V2I components of CV technologies.

The V2I hub sends the detector data to a physical traffic signal controller via National Transportation Communications for ITS Protocol (NTCIP), which then generates corresponding phase status data and sends it back to the V2I hub. Next, the V2I hub sends real-time phase status to VISSIM via the HIL interface; the phase status of a virtual signal controller in VISSIM is synchronized with the phase status of the physical traffic signal controller in real time.

Additionally, the V2I hub encodes the virtual vehicle BSMs, SPaT messages, and MAP messages and sends them to an RSU. The RSU transmits these DSRC messages over the air. The physical vehicle in the field is equipped with an OBU, and the OBU receives and decodes the BSMs, SPaT messages, and MAP messages from the RSU. At the same time, the OBU of the physical vehicle transmits BSMs at 10 Hz over the air.

There are two ways to synchronize physical vehicles into VISSIM. One method involves the RSU receiving these BSMs from the physical vehicles and sending them to the V2I hub, which then decodes the messages and sends the status data of the physical vehicle to the HIL interface. The HIL interface passes this data back to VISSIM. The real-time location information of the physical vehicle is decoded from the BSM of the physical vehicle, while the position of the virtual vehicle is extracted from the COM interface in VISSIM; both the physical and virtual vehicle information is obtained at a frequency of 10 Hz. An alternative method for sending the location of the physical vehicles to the HIL interface is via ethernet; this approach avoids potential delay caused by message communication over the air without significant loss in realism. This approach requires wireless network coverage at the test site. In the meantime, BSMs from virtual vehicles still need to be transmitted over the air to replicate real-world conditions. This latter approach is preferred since it has been proven by initial testing that it significantly reduces synchronization delay—down to less than 20 ms, as compared with above 100 ms in the former approach (at the test site, there are many trees that may affect DSRC communication). However, the former approach can be used when no wireless connection is available at a test site.

In this study, high-accuracy GPS data was provided by PinPoint, an enhanced GPS device that fuses data from multiple sensors, including GPS, to create far more reliable positioning, velocity, orientation, and time metrics. This suite of sensors is always active, weighing all data to ensure accurate positioning even if GPS data is unavailable or inaccurate. PinPoint has a small relative location error (less than 1 m) and generates high frequency data (50 Hz).

Vehicle status data can be directly obtained from the vehicle's CAN bus (e.g., current speed, acceleration rate, and deceleration rate) and used for preparing BSMs. By using BSM data every 0.1 second, the virtual vehicle, which represents the physical vehicle in the field, has a smooth trajectory when synchronized into the VISSIM simulation. More importantly, accurate and smooth vehicle trajectory and dynamic profiles are critical for measuring the effectiveness of a CAV application.

The advanced applications can be implemented in the onboard computer of a physical vehicle. The necessary real-time inputs for the application are obtained from in-vehicle data (e.g., vehicle status data), data broadcasted by RSUs (i.e., SPaT), and other virtual vehicle data sent from VISSM through DSRC. The application automatically calculates recommended vehicle control commands. These commands are then directly injected into the physical vehicle lower-level control via a research CAN bus.

Selected Testing Facility

Turner-Fairbank Highway Research Center (TFHRC) was identified as the test track for the proposed HIL system. This facility was selected because all the necessary equipment is available there from previous CAV projects. Several RSUs are installed at the testbed. One traffic signal controller that is compatible with NTCIP standards is used. A V2I hub computer is configured and connected to the intersection to generate MAP messages based on the prepared MAP data. The MAP data for the intersection at TFHRC is prepared according to previous surveying data at the intersection. The RSUs, traffic signal controller, and the V2I hub are connected to the SDN network, a research ethernet network developed for research purposes at TFHRC.

One CAV—a Cadillac SRX—was used in field tests. The vehicle is equipped with long-range forward-looking radar, in-vehicle Linux PC, corrected GPS, OBU, and MicroAutoBox (MAB). Figure 2 demonstrates the data flow between different vehicle system components. At the center of the vehicle control system is the in-vehicle Linux PC. PinPoint transmits real-time high-accuracy GPS data to the in-vehicle computer. The OBU broadcasts BSMs, receives other vehicles' BSMs, and transmits this information to the Linux PC. The long-range radar transmits object data to the Linux PC. The MAB receives data that is processed by the Linux PC, including BSMs from other vehicles, RSUs, radar data, and processed data (for some applications, part of the algorithm is implemented in the Linux PC). The MAB then processes this information and outputs a final speed command that is seamlessly injected into the vehicle CAN bus. The testing CAV was also recently equipped with an advanced fuel meter that can measure fuel consumption of vehicle operations in real time.

Queue-Aware Signalized Intersection Approach and Departure

This paper applies the developed HIL testing system to evaluate one example CAV application, the Q-SIAD

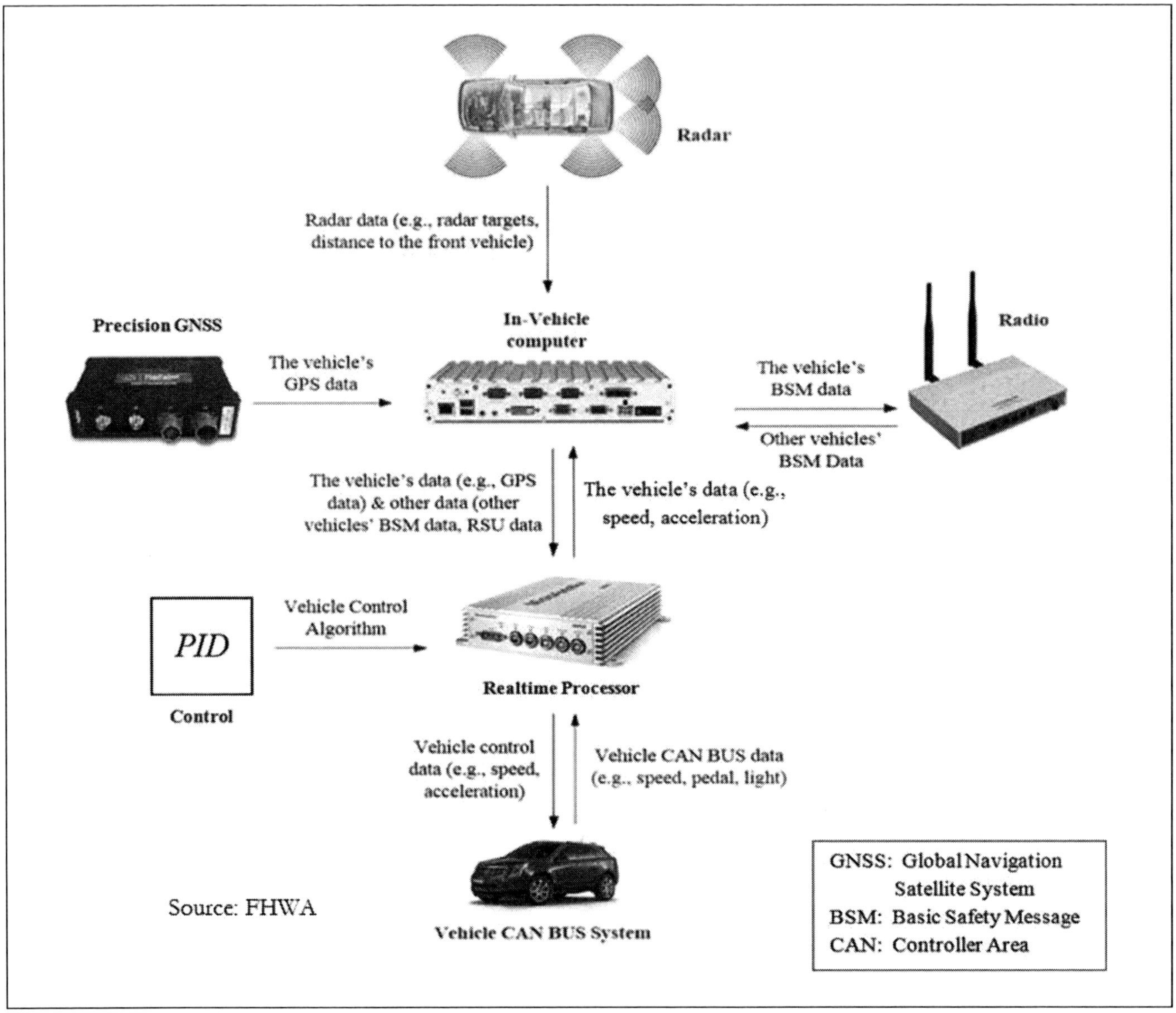

Figure 2. Data flow of the vehicle control systems.

application, which aims at minimizing fuel consumption and emissions, increasing intersection throughput, and enhancing driving comfort. The application generates recommended speed profiles for testing CAVs based on the vehicle's status (e.g., current speed), SPaT and MAP information, system constraints, and parameters (e.g., maximum acceleration) (*13, 14*).

Most existing SIAD evaluation studies are limited to simulation experiments and obtained for limited hypothetical environments. A small number of field experiments (e.g., (*15*)) limit themselves to a single SIAD vehicle without considering the existence of other vehicles in the traffic stream, which can have significant effects on its effectiveness from the following two aspects:

Background Traffic and Queue Length: The subject CAVs "optimal" SIAD trajectory may have conflicts with other vehicles' desired trajectories. CAVs in high density streams may be limited in adjusting their speed, especially speeding up, when they are required to follow non-CAVs. The existing queue length at the intersection is another critical factor, potentially forcing the SIAD-equipped vehicles to decelerate or stop; therefore, having a high impact on the fuel consumption and efficiency.

Existence of CVs: Within a CV environment, vehicles equipped with communication devices can provide their status information to the nearby infrastructure and CAVs in real time. These data can be used to estimate the surrounding traffic condition and intersection queue length.

Algorithm Framework

Figure 3*a* shows the framework of the proposed algorithm. The subject CAV checks at every time step if a front vehicle is nearby. If yes, the CAV needs to switch to linear approaching mode and then car-following mode to follow the front vehicle. If there is no front vehicle, the Q-SIAD algorithm uses CV data from the traffic to estimate the current queue length and consider a safety buffer distance upstream of the end of the last vehicle in the queue as a virtual "stop bar." The purpose is to let the subject CAV plan the trajectory better and avoid a complete stop at the end of the queue. When the estimated queue length is different from that of the previous time step, the algorithm needs to be applied again to generate updated speed profiles for the subject vehicle.

The linear approaching mode is defined in Equation 1.

$$v_r = \frac{v_r^s(d_r - d_r^e)}{d_r^s - d_r^e} \tag{1}$$

where v_r and d_r are the relative speed and distance between the subject CAV and front vehicle, respectively; v_r^s and d_r^s are the relative speed and distance at the beginning of the linear approaching mode, respectively; d_r^s is the threshold that triggers the switch from the free driving mode to linear approaching mode; v_r^e indicates the relative speed at the end of the linear approaching mode; and d_r^e is the threshold that triggers the switch from the linear approaching mode to car-following mode. Theoretically, $v_r^e = 0$ is the start of the car-following mode.

The car-following mode is governed by a gap regulator. A simplified gap regulator (*16*) is used in the proposed system for testing purposes. Equations 2 and 3 show the calculation of a gap error and subject vehicle speed at each time step.

$$e_n = x_{n-1} - x_n - t_{hw}v_n \tag{2}$$

$$v_n = v_n^{prev} + k_1e_k + k_2\dot{e}_k \tag{3}$$

where e_n is the gap error for the subject vehicle; x_{n-1} and x_n are location of the front and subject vehicle, respectively; t_{hw} is the desired time gap setting; v_n and v_n^{prev} are subject vehicle speeds in the current and previous time steps, respectively; $k_1 = 0.45$ and $k_2 = 0.25$ are control gains.

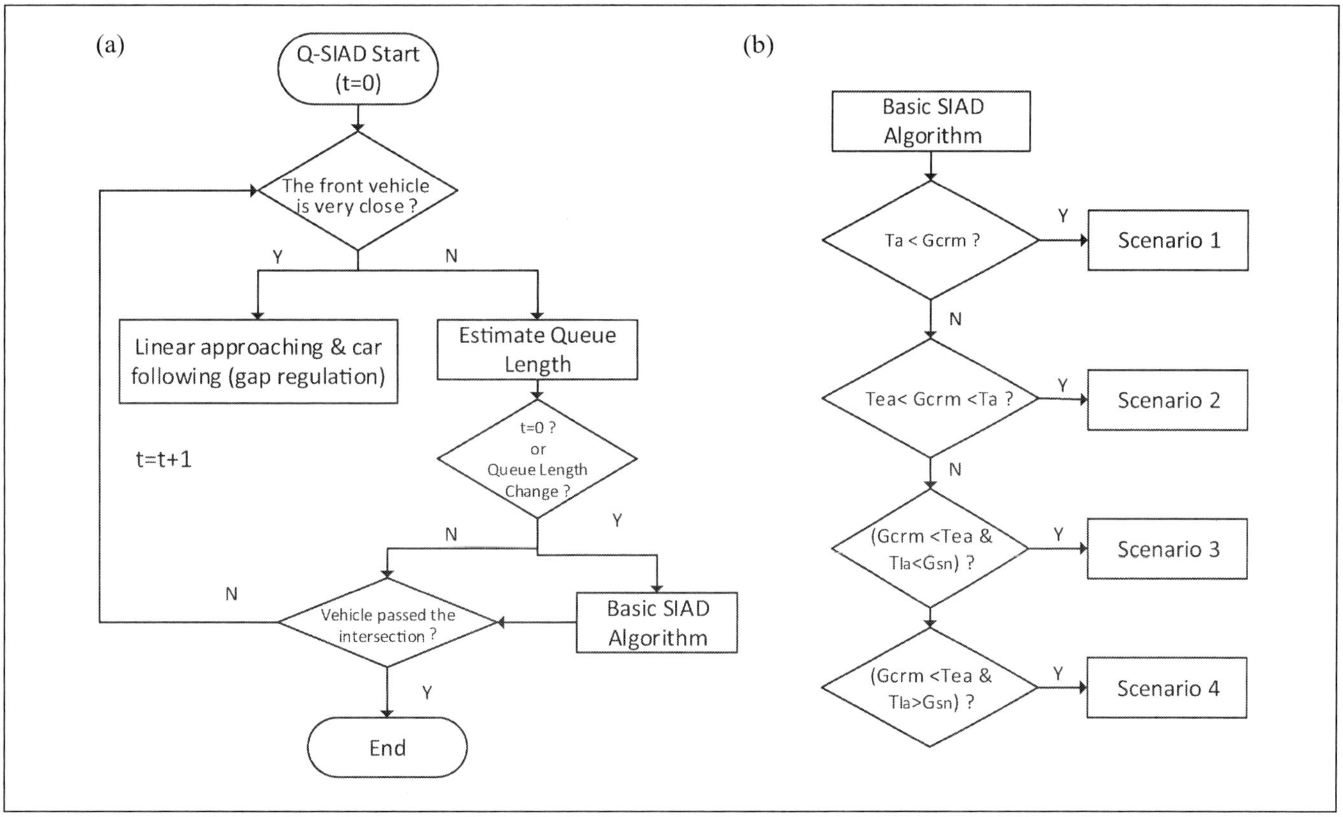

Figure 3. Framework of the queue-aware signalized intersection approach and departure algorithm.

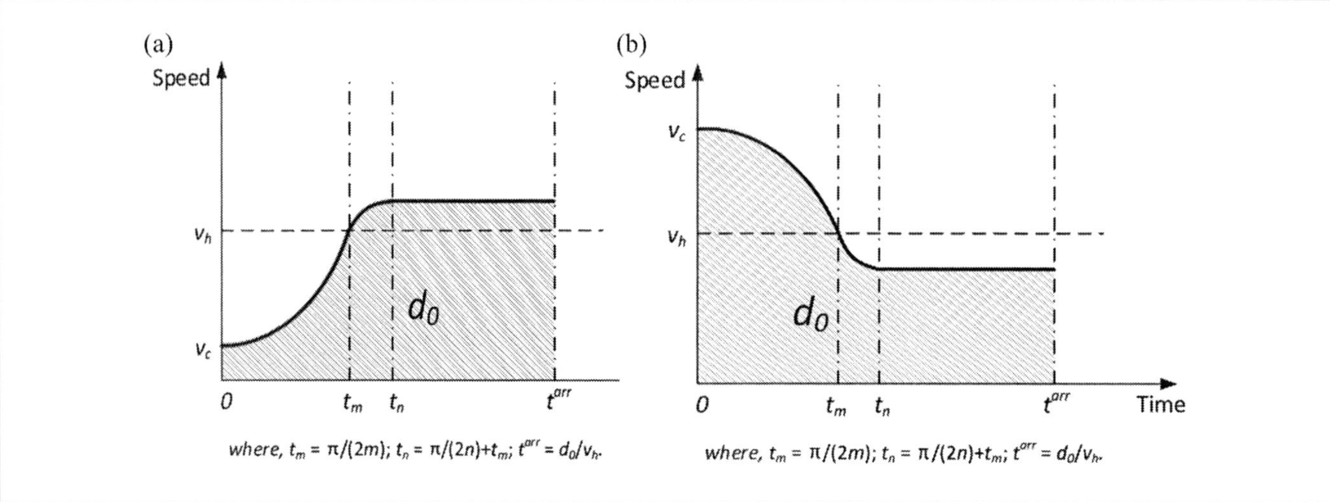

Figure 4. Acceleration (*a*) and deceleration (*b*) profiles of the VTPA (*17*).

One of the core components of the Q-SIAD algorithm is labeled "Basic SIAD Algorithm" in Figure 3*a*. This algorithm was developed and applied in a previous field experiment, and entitled the vehicle trajectory planning algorithm (VTPA) (*15*). VTPA has three major inputs: (a) the vehicle's status; (b) DSRC messages from RSUs, such as SPaT and MAP messages; and (c) system constraints and parameters. VTPA obtains current remaining green time (if the light is green at the time) or current remaining time until the next green light (if the light is red). VTPA also estimates a major parameter, distance from the subject vehicle to the stop bar, at each time step. After that, VTPA calculates the earliest and latest times that a testing vehicle would arrive at the stop bar, according to system constraints and parameters.

Figure 3*b* shows the flow chart of four possible scenarios for SIAD trajectory design. In Scenario 1, when the predicted arrival time of the testing vehicle at the stop bar T_a is less than the remaining time for the current green light G_{crm}, the subject vehicle can pass the target intersection by keeping its current speed. In Scenario 2, if $T_a < G_{crm}$ and the predicted earliest arrival time of the subject vehicle at the stop bar T_{ea} is less than G_{crm}, then the subject vehicle can pass the target intersection by speeding up in an eco-friendly manner. In Scenario 3, if $T_{ea} < G_{crm}$ and the predicted latest arrival time of the subject vehicle at the stop bar T_{la} is earlier than the beginning of the next green light G_{sn}, then the subject vehicle has to stop. Finally, in Scenario 4, when $T_{ea} < G_{crm}$ and $T_{la} > G_{sn}$, the subject vehicle needs to slow down and pass the intersection at the beginning of the next green light. In addition, the algorithm determines whether the subject vehicle has passed the target intersection. If it has, the algorithm ends or considers the next intersection.

After corresponding scenarios are identified, the VTPA trajectory generator produces speed profiles for the subject vehicle. The objective of the basic SIAD algorithm is to minimize acceleration and deceleration and maintain the initial speed as much as possible. Specifically, the piecewise trigonometric-linear function is selected to generate speed profiles. Figure 4 shows acceleration profiles and deceleration profiles generated by the piecewise trigonometric-linear function (where t^{arr}: time of arrival; v_c: the current speed; m, n: parameters of the piecewise trigonometric-linear function; d_0: the distance to stop bar). For details about piecewise trigonometric-linear functions and generating speed profiles for each scenario, please refer to Leidos (*15*) and Xia (*13*).

Note that in the Q-SIAD application, real-time BSM data from connected vehicles is collected. If a queue is detected at the downstream signalized intersection, the position of the rear bumper of the last vehicle with an extremely low speed is considered as the position of the end of the queue. As shown in Figure 3*a*, when the estimated queue length changes, the basic SIAD algorithm is applied again to generate updated speed profiles for the subject vehicle. When the penetration rate of CVs is 100%, the estimated queue length is accurate. However, when the penetration rate of connected vehicles is less than 100%, the estimated queue length may be shorter than the actual queue length. When the algorithm does not estimate queue length using CV data, it is reduced to the basic SIAD algorithm.

Experimental Design and Results Analysis

Experimental Design and Execution

Two experimental cases were tested and compared. The baseline without Q-SIAD used production adaptive cruise control modified with the capability of stopping at a red light and following preceding virtual vehicles. Real-time gap information between the CAV and the vehicle

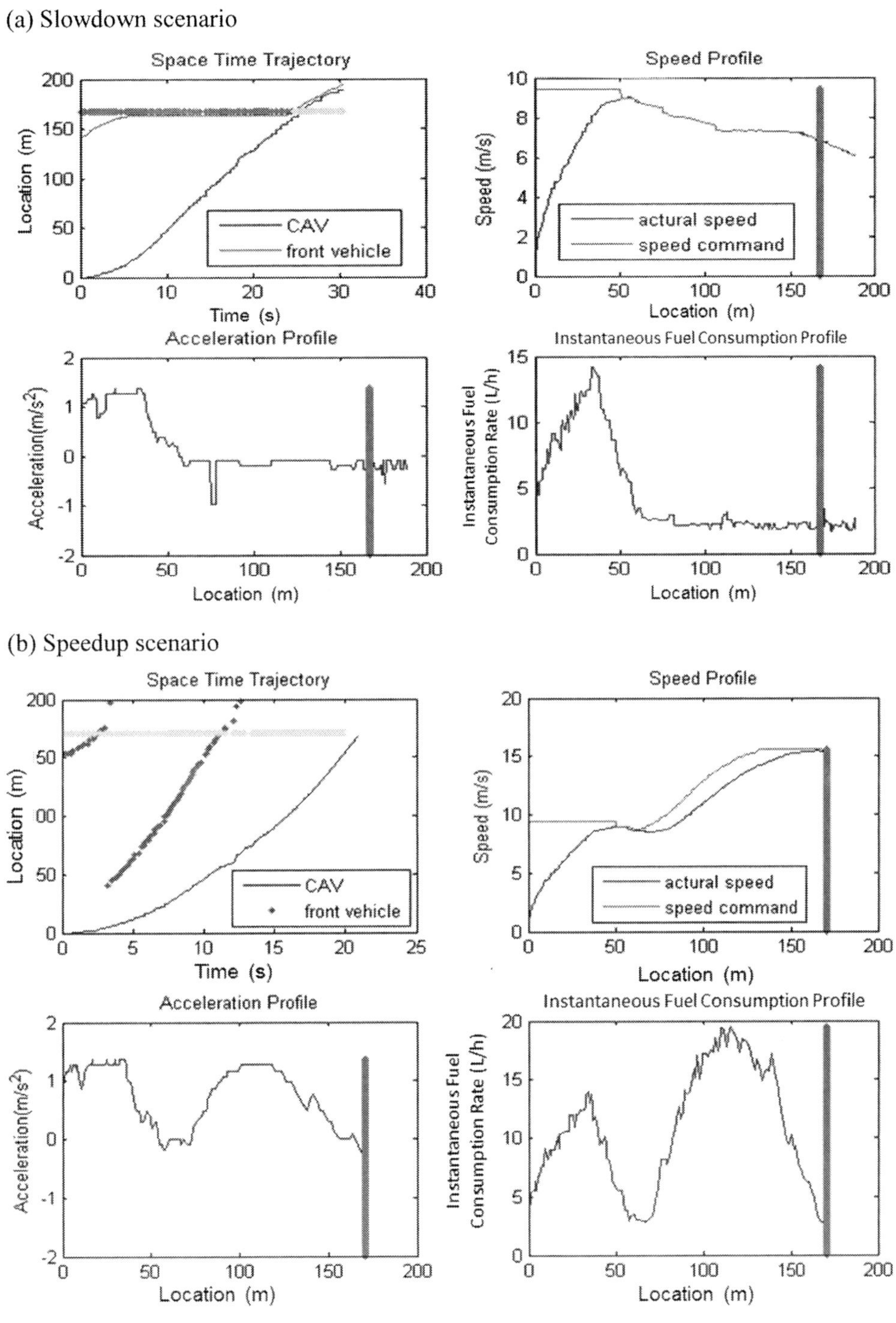

Figure 5. Example data visualization for different scenarios.

which is immediately in front of the CAV was extracted from VISSIM and injected to the vehicle platform. The test CAV started from about 175 meters upstream of the intersection, with a constant set speed of 20 mph, and the test ended 50 meters downstream of the intersection. The test CAV either accelerated to the set speed if no preceding vehicle was detected (virtually) or accelerated to follow the preceding vehicle if a front vehicle was detected. The test vehicle proceeded to the end of the experimental segment without causing safety concerns, that is, safety conflicts with simulated vehicles.

Table 1. Aggregate Fuel Consumption (in Milliliters) for Different Scenarios

Start time stamp	SIAD scenario	Baseline	Basic SIAD		Queue-aware SIAD			
					0% CV		100% CV	
R0	Slowdown	43.6	19.1	56.2%[a]	19.3	55.7%[a]	19.3	0.0%[b]
R5	Slowdown	43.6	19.1	56.2%	20.5	53.0%	20.3	1.0%
R10	Cruise	43.6	20.5	53.0%	20.5	53.0%	20.1	2.0%
R15	Cruise	22.5	21.4	4.9%	24.2	−7.6%	23.1	4.6%
R20	Cruise	22.5	21.4	4.9%	20.9	7.1%	20.9	0.0%
G0	Cruise	22.5	20.8	7.6%	22.4	0.4%	22.9	−2.2%
G5	Speedup	43.6	33.2	23.9%	39.1	10.3%	38.1	2.6%
G10	Stop	43.6	42.5	2.5%	43.4	0.5%	42.7	1.6%
G25	Stop	43.6	42.3	3.0%	43.5	0.2%	43.5	0.0%

[a]Percent change using the corresponding baseline scenario as the base case.
[b]Percent change using the corresponding 0% MP scenario as the base case.

In the Q-SIAD case, the test intersection was controlled by the same fixed-timing signal. The test CAV received a recommended speed profile that took into account the impacts of front vehicles and followed the profile automatically while approaching the test intersection. This scenario was tested with variations of different system parameters: (a) traffic density: either no traffic and medium traffic of 300 vphpl (usually 2–3 vehicles were in front of the test vehicle and before the stop bar); (b) CV market penetration (0% and 100%); and (c) different vehicle start times throughout the entire traffic signal cycle (i.e., every 5 seconds in the 60-second cycle) for each different run. The collected data, such as trajectories of physical and virtual vehicles and fuel consumption of the physical vehicle via the fuel meter, were used to analyze fuel consumption savings.

During the field experiment, each scenario was tested three times. The difference from run to run was minor in this experiment because generated medium traffic was similar and usually only one to three vehicles were in front of the subject vehicle and before the stop bar. The subject vehicles accelerated to the operating speed of 20 mph and then started trajectory control. Data were collected until around 50 meters downstream of the stop bar after the vehicle resumed the operating speed.

Experimental Results

Figure 5*a* and *b* depict visualizations of data collected from two experimental runs using four types of plots: space-time trajectory, profiles of speed, acceleration, and instantaneous fuel consumption.

Figure 5*a* demonstrates a test vehicle slowdown scenario when one CV stops at the stop bar. The space-time trajectory shows that the subject vehicle slows down and passes the intersection right after the CV passes the intersection. Without the queue-aware components of the algorithm, it is expected that the subject vehicle would arrive at the intersection earlier and it is likely the vehicle may be forced to stop completely. The speed and acceleration profiles show that the vehicle slows down in the beginning and successfully passes the intersection without a stop. The fuel consumption profile shows that the optimized trajectory makes it possible for the vehicle to pass the intersection with very little fuel use, since no stopping is involved.

Figure 5*b* demonstrates a subject vehicle speedup scenario where the vehicle passes the intersection right before the red time starts. The first three subplots show that the subject vehicle calculates a trajectory that accelerates the vehicle to a certain level and then smoothly passes the intersection. The algorithm is also proven effective in terms of planning and executing vehicle trajectory relatively accurately. The fuel consumption plot shows a relatively large instantaneous fuel consumption rate during the speedup process. Therefore, a calculation of fuel consumption value is required to compare the total fuel consumption of a stop scenario and a speedup scenario. In the interests of space, visualizations of cruise and stop scenarios are not shown in this paper. Fuel consumption in these two types of scenarios is generally similar.

The actual data used to calculate the fuel consumption is from the location where the vehicle's speed first accelerates to 20 mph to the end of the data collection (50 meters after the stop bar). This is to ensure fair comparison between different runs on how much fuel is needed to pass the intersection segment after the vehicle resumes its original operating speed. The total segment for data analysis is about 170 meters.

Table 1 summarizes average fuel consumption results and benefits under different scenarios. For all scenarios, the traffic controller had a single fixed time of 30 seconds of red, 27 seconds of green, and 3 seconds of yellow. The first column indicates the start time of the scenario with regard to the signal timing. The baseline column shows the cruise control case with background traffic (and stops at red lights). The basic SIAD column shows the level of fuel consumption when no background traffic exists, and these values provide potential "upper bounds" of benefits of Q-SAID. Q-SIAD column shows results with background traffic. The 0% market penetration and 100% market penetration columns reflect the results when no CV data and all CV data for background traffic are available to the Q-SIAD algorithm.

In Figure 5*a*, the fuel consumption in the slowdown scenarios is significantly lower than the baseline (i.e., complete stop). For instance, in Table 1, the benefits of R0 scenario, for which the vehicle starts at red signal at time zero, is fuel savings of over 55%. The benefits of Q-SIAD with background traffic are quite close to the ideal benefit of basic SIAD for no background traffic conditions. For speedup scenarios, according to Table 1, the ideal benefit is more than 20% while the existence of background traffic reduces the number to around 10–13%, which is still quite a significant saving compared with the baseline.

In Figure 5*b*, some test runs have a moving front vehicle that blocks the subject vehicle, which cannot accelerate to the target speed. Therefore, the subject vehicle first accelerates and then decelerates to follow the front vehicle. For some runs of this scenario, the subject vehicle needed to stop completely, depending on how much the front vehicle's movements impacted the subject vehicle. The number presented in the Table 1 is averaged from three runs. However, even though extra fuel may be consumed when speeding up (as shown in Figure 5*b*), on average great benefits still can be seen for speedup scenarios.

In Table 1, cruise scenario R10, for which the vehicle starts at red signal timing of 10 seconds, is unique because the vehicle needs to decelerate and completely stop in the baseline scenario. This is because the signal is still red when the vehicle is near the stop bar. However, the vehicle can cruise and is able to pass the intersection at the beginning of green. This shows the significance of transmitting SPaT messages to vehicles and using an effective Q-SIAD algorithm. For other cruise scenarios, the benefits are relatively small because in both baseline and Q-SIAD scenarios the subject vehicle cruises throughout the experimental segment. The minor differences between runs are because of the impact of slightly different background traffic. For stop scenarios, the subject vehicle needs to come to a complete stop for both the baseline and the Q-SIAD algorithm, and thus the fuel consumption benefit is also limited.

It is also vital to mention that after the establishment of this HIL system, it is an iterative process to test whether the Q-SIAD software can make the application function. Many parameters, such as safety buffers to prevent red light running and vehicle response delays, need to be added to ensure safe operation of automated signalized intersection approach and departure. In addition to quantifying benefits as tabulated in Table 1, testing Q-SIAD application software (and hardware) to make the system work is a major use case for such HIL systems.

Conclusions and Future Research

Using emerging HIL testing allows physical test vehicles to interact with virtual vehicles from traffic simulation models, providing an evaluation environment that can replicate actual deployment conditions at early stages of CAV development without incurring excessive costs. In this study, the hardware involved included a physical CAV controlled in real time, a physical traffic signal controller, and DSRC devices (e.g., OBU, RSU). The VISSIM traffic simulation tool was used to set up the simulation environment and generate virtual background traffic. Using HIL systems with physical vehicles offers many benefits; for example, such systems increase the validity of findings by taking into account the physical vehicle's trajectory in a simulated traffic environment. In addition, the possibility of collecting data on various CAV applications in different traffic scenarios enables better calibration of CAV driver models, which can serve as the foundation for large-scale evaluations with corresponding SIL systems.

The HIL system developed was used to test a representative early deployment application of CAV, queue-aware signalized intersection approach and departure (Q-SIAD), which aims to minimize fuel consumption and emissions, increase intersection throughput, and enhance driving comfort. The algorithm generated recommended speed profiles for testing CAVs based on a vehicle's status (e.g., location, current speed), SPaT and MAP information, and system constraints and parameters (e.g., maximum acceleration and deceleration). The algorithm also considered the status of other vehicles (i.e., virtual connected or virtual non-connected vehicles) in designing trajectories for the CAV. The experiment was conducted in the simulated environment of the test track at the TFHRC. It successfully demonstrated this functionality with one physical testing CAV (and other virtual vehicles) driving through one intersection controlled by a fixed-timing traffic signal.

There are still limitations of the developed HIL system that need to be addressed in future studies. First, this platform is developed for testing CAV V2I applications. Modifications of the system are required to test V2V applications—particularly safety critical applications—where the performance requirement of system synchronization between the virtual and physical worlds needs to be enhanced. Second, this study conducted testing at a small-scale, low-speed site. Future studies should consider more realistic testing scenarios, including high-speed traffic or complex vehicular movements, on signalized arterials. Third, other hardware and software components need to be incorporated when evaluating different applications. For example, when the driver behavior is of interest, driving simulators or immersive technologies (e.g., virtual reality, augmented reality) can be included.

Acknowledgments

This study is funded by the US Department of Transportation (Project Number: DTFH61-12-D-00030). The work was conducted by the Leidos team using the vehicle fleet and connected vehicle facility at the FHWA's Turner-Fairbank Highway Research Center.

Author Contributions

The authors confirm contributions to the paper as follows: study conception and design: Jiaqi Ma, Zhitong Huang, Fang Zhou, Christopher L. Melson, and Rachel James; Data collection: Jiaqi Ma, Fang Zhou, Zhitong Huang, and Xiaoxiao Zhang; Analysis and interpretation of results: Jiaqi Ma, Fang Zhou, Zhitong Huang. Draft manuscript preparation: all authors prepared the draft manuscript. All authors reviewed the results and approved the final version of the manuscript.

References

1. Volvo Car Group. *Volvo Cars Presents a Unique Solution for Integrating Self-Driving Cars into Real Traffic*. February 2015. https://www.media.volvocars.com/global/en-gb/media/pressreleases/158276/volvo-cars-presents-a-unique-system-solution-for-integrating-self-driving-cars-into-real-traffic. Accessed June 14, 2016.
2. Jaynes, N. *Volvo Plans to Test 100 Autonomous Cars on Public Roads in China*. Mashable, April 2016. http://mashable.com/2016/04/06/volvo-autonomous-cars-china/#8TvojVHGAmqW. Accessed June 14, 2016.
3. Intelligent Transportation Systems Joint Program Office. *Connected Vehicle Pilot Deployment Program*. 2016. http://www.its.dot.gov/pilots/index.htm. Accessed June 14, 2016.
4. Fathy, H. K., Z. S. Filipi, J. Hagena, and J. L. Stein. Review of Hardware-in-the-Loop Simulation and its Prospects in the Automotive Area. In *Defense and Security Symposium; vol. 6228 of SPIE Proceedings*. SPIE, 2006, pp. 1–20.
5. Bullock, D., J. M. Morales, and B. Sanderson. Impact of Signal Preemption on the Operation of the Virginia Route 7 Corridor. *Proc., 1999 ITS America Conference*, Washington D.C., 1999.
6. So, J., J. Lee, and B. Park. Evaluation of Manual Traffic-Signal Control under Oversaturated Conditions Using Hardware-in-the-Loop Simulation. *Journal of Transportation Engineering*, Vol. 139, No. 11, 2013, pp. 1068–1075.
7. Sanchez, R. O., R. Horowitz, and P. Varaiya. Hardware-in-the-Loop On-ramp Simulation Tool to Debug and Test the Universal Ramp Metering Software. *IFAC Proceedings Volumes*, Vol. 42, No. 15, 2009, pp. 211–216.
8. Klüver, M., C. Herrigel, C. Heinrich, H.-P. Schöner, and H. Hecht. The Behavioral Validity of Dual-task Driving Performance in Fixed and Moving Base Driving Simulators. *Transportation Research Part F: Psychology and Behaviour*, Vol. 37, 2016, pp. 78–96.
9. Regan, M., J. D. Lee, and K. L. Young. *Driver Distraction: Theory, Effects, and Mitigation*. CRC Press, Boca Raton, Fla., 2008.
10. Zulkefli, M. A. M., P. Mukherjee, Z. Sun, J. Zheng, H. X. Liu, and P. Huang. Hardware-in-the-Loop Testbed for Evaluating Connected Vehicle Applications. *Transportation Research Part C: Emerging Technologies*, Vol. 78, 2017, pp. 50–62.
11. Hou, Y., Y. Zhao, K. F. Hulme, S. Huang, Y. Yang, A. W. Sadek, and C. Qiao. An Integrated Traffic-driving Simulation Framework: Design, Implementation, and Validation. *Transportation Research Part C: Emerging Technologies*, Vol. 45, 2014, pp. 138–153.
12. The Texas A&M Transportation Institute, Siemens, and Battelle. *New Approaches for Testing Connected Highway and Vehicle Systems Design Document (Draft)*. Publication FHWA BAA No. DTFH61-13-R-00011. Federal Highway Administration, U.S. Department of Transportation, Washington, D.C., 2014.
13. Xia, H. *Eco-Approach and Departure Techniques for Connected Vehicles at Signalized Traffic Intersections*. PhD dissertation. University of California, 2014. http://escholarship.org/uc/item/5xz408r4.
14. Ma, J., X. Li, F. Zhou, J. Huc, and B. Park. Parsimonious Shooting Heuristic for Trajectory Design of Connected Automated Traffic Part II: Computational Issues and Optimization. *Transportation Research Part B: Methodological*, Vol. 95, 2017, pp. 421–441. http://dx.doi.org/10.1016/j.trb.2016.06.010.
15. Leidos Inc. *TO17: GlidePath Prototype Application*. Final Report. Federal Highway Administration, U.S. Department of Transportation, Washington D.C., 2016.
16. Milanés, V., and S. E. Shladover. Modeling Cooperative and Autonomous Adaptive Cruise Control Dynamic Responses Using Experimental Data. *Transportation Research Part C: Emerging Technologies*, Vol. 48, 2014, pp. 285–300.

The Standing Committee on Vehicle-Highway Automation (AHB30) peer-reviewed this paper (18-05431).

The work presented in this paper remains the sole responsibility of the authors.

TRR
JOURNAL OF THE TRANSPORTATION RESEARCH BOARD

Article

Transportation Research Record
2018, Vol. 2672(22) 47–57

DOI: 10.1177/0361198118797831
journals.sagepub.com/home/trr
SAGE

Safety Evaluation of the Advanced Stop Assist System in Connected Vehicle Environment

Gwamaka Njobelo[1], Thobias Sando[1], Soheil Sajjadi[2], Enock Mtoi[3], Eren Erman Ozguven[4], and John Sobanjo[4]

Abstract

Although traffic signals are installed to reduce the overall number of collisions at intersections, certain types, in particular, rear-end collisions are increasing due to signalization. One dominant factor associated with rear-end crashes is the indecisiveness of the driver, especially in the dilemma zone. An advisory system to help the driver make the stop-or-pass decision would greatly improve intersection safety. This study proposes and evaluates an Advanced Stop Assist System (ASAS) at signalized intersections by using Vehicle-to-Infrastructure (V2I) communication. The proposed system utilizes communication data, received from roadside equipment, to provide approaching vehicles with vehicle-specific advisory speed messages to prevent vehicle hard-braking at a yellow or red signal. A simulation test bed was modeled using VISSIM, a microscopic simulation software, to evaluate the effectiveness of the proposed system. The results demonstrate that at full market penetration (100% saturation of vehicles equipped with on-board communication equipment), the proposed system reduces the number of hard-braking vehicles by nearly 50%. Sensitivity analyses of market penetration rates also show a degradation in safety conditions at penetration rates lower than 40%. The results suggest that a penetration rate of at least 60% is required for the proposed system to minimize rear-end collisions and improve safety at the signalized intersections.

The American Association of State Highway and Transportation Officials (AASHTO) offered a nationwide challenge to deploy Dedicated Short Range Communications (DSRC) infrastructure with Signal Phase and Timing (SPaT) broadcast on at least one corridor (approximately 20 signalized intersections) in each of the 50 states by January 2020 (*1*). These corridors are expected to play a significant role as test beds for implementing innovative initiatives in the connected and autonomous vehicle technology. One of the potential uses of such deployment would be an implementation of an intersection crash avoidance system based on Vehicle-to-Infrastructure (V2I) communicaiton. Signalized intersections are associated with different types of conflicts due to various opposing movements and differential speeds. Although traffic signals are installed to reduce certain types of collisions (e.g., head-on and angle crashes), they are known to increase rear-end collisions (*2*, *3*) . In 2013, a total of 4,700 fatal crashes, 826,000 injury crashes, and 1.76 million property-damage-only crashes occurred in relation to signalized intersections in the United States; nearly 30% of these were rear-end collisions (*4*). Rear-end crashes at intersections are caused in part by differential speeds as vehicles approach the intersection, followed by abrupt decelerations at the onset of yellow and red signals. The indecision of drivers in these situations is likely to result in harder braking, thus increasing the likelihood of a rear-end crash.

Currently, there are several in-vehicle systems that use V2I communication to provide information of the signal states to drivers. For example, in Tampa, Florida, there are two V2I systems that are currently being beta-tested. The first system, *EnLighten*, developed by Connected Signals (*5*), is a cloud based V2I system that uses a smartphone application to show the amount of remaining green time and if the driver is stopped at a red signal the system advises the driver of the remaining amount of red

[1]School of Engineering, College of Computing, Engineering and Construction, University of North Florida, Jacksonville, FL
[2]Arcadis G&M of North Carolina Inc., Raleigh, NC
[3]AECOM Florida's Turnpike, Milepost, Pompano Beach, FL
[4]Department of Civil and Environmental Engineering, FAMU-FSU College of Engineering, Tallahassee, FL

Corresponding Author:
Address correspondence to Gwamaka Njobelo: gnjobelo@gmail.com

time before the next green phase. In the second system, a U.S. Department of Transportation (DOT) pilot project, Siemens partnered with the Tampa-Hillsborough Expressway Authority (THEA) to install a V2I system in downtown Tampa to provide an advisory speed to drivers who wish to travel without stopping before the next green phase (*6*). Whereas *EnLighten* is based on analytics of the signal grid master data, the second system uses DSRC and requires communication devices to be installed at both the intersection and in the vehicle to enable short range communication. The V2I technology has the potential to reduce rear-end conflicts at signalized intersections.

There is a need to devise a system referred to here as the Advanced Stop Assist System (ASAS), which takes advantage of V2I communication technology to notify drivers to prepare to stop when the remaining green time is insufficient to clear the intersection under prevailing traffic conditions. Since the CV technology allows vehicles to communicate with each other and with the infrastructure, the exchange of information can be used to provide an advisory speed message in advance. It is worth pointing out that messages are sent not only to vehicles in the dilemma zone, but also to any qualifying vehicle that is within the communication range. Each vehicle will receive an advisory message when the signal is still green. Earlier stopping maneuvers would allow for smooth deceleration and potentially reduce the number of vehicles trapped in the dilemma zone, therefore improving safety by reducing rear-end collisions and red-light running, a leading cause of angle crashes.

Objectives

The objectives of this study are: (1) to propose and demonstrate the implementation of ASAS in a microscopic simulator; and (2) to evaluate the potential safety benefits of ASAS. In particular, the reduction of rear-end collisions under the CV environment with V2I communications. Thus, this manuscript presents an algorithm used to develop the ASAS under the CV environment in the microscopic traffic simulator. The effectiveness of the proposed ASAS algorithm is assessed using two performance measures: speed profiles as vehicles approach the intersection to examine speed smoothness, and maximum deceleration rates to determine the effectiveness of the system in reducing hard-braking. The proposed system was implemented and evaluated using the Car-to-Devices (C2X) simulation module available in the VISSIM microscopic simulation software.

Literature Review

There are numerous research efforts that have documented the Connected Vehicle (CV) technology and its potential to improve the mobility, safety, and environmental sustainability of transportation systems. The following sections provide a summary of the CV literature pertinent to this study.

CV Technology

The history of CV can be traced back to 2003 when the U.S. DOT first launched the Vehicle-Infrastructure-Integration (VII) program (*7*). The initial objective of VII was to address the traffic safety problems through high-speed wireless communications among Vehicle-to-Vehicle (V2V) and V2I. The CV framework is made up of three critical elements; On-Board Equipment (OBE), Roadside Equipment (RSE), and Back-Office Servers (*8*). OBE consists of devices embedded in the vehicle that support DSRC with nearby vehicles and RSE. RSE consists of roadside devices that support DSRC with nearby OBE-equipped vehicles within the communication distance, other RSEs, and the control centers. Back-office servers represent the control center that connects RSEs and monitors the traffic network.

The Federal Communications Commission (FCC) has allocated the spectrum from 5.850 to 5.925 GHz, that is, the "5.9 GHz band", for DSRC operations in the United States. This spectrum is divided into seven 10 MHz channels with a 5 MHz guard (reserve) at the low end. Pairs of 10 MHz channels can also be combined into a 20 MHz channel (*9*). Most of the wireless devices participating in safety applications have a maximum output power of 20 dBm and a communication zone of nearly 400 m (*9*). Due to hidden terminal problems and multipath interference, the effective range of communication may be lower than the nominal value. Where now vehicles collect and convey a lot of information, just a few years ago vehicles relied completely on low technology methods to communicate with each other and the environment by using turn signals, brake lights, and static signs. In recent years, technology that provides wireless communication between vehicles and transportation infrastructure has increased. This technology is a combination of a number of technological advancements including advanced wireless communications, on-board computer processing, advanced sensors, Global Positioning System (GPS) navigation and smart infrastructure to provide a networked environment (*10*). A CV environment allows high-speed exchange of information between vehicles and infrastructure. Signal systems can use data from in-vehicle sensors transmitted wirelessly from equipped vehicles to the signal controller. Several measures can be accessible such as vehicle speeds, positions, arrival rates, acceleration or deceleration rates, stopped time, and queue lengths (*11*).

CV at Intersections

Several studies have examined the use of CV for intersection control. Many of these studies focused on the optimization of signal phases by taking advantage of the information sent from the equipped vehicles. Guler et al. (*12*) developed a traffic control algorithm to minimize delay using information from CV to mitigate urban congestion. V2I safety applications were proposed to address problems that V2V communications would not address (*13*). Safety applications that use V2I only require RSE at targeted facilities, normally at intersections. There are V2I safety applications that have been developed at intersections and tested for their effectiveness through simulation studies and field test beds. Some of these applications include Cooperative Intersection Collision Avoidance System (CICAS) framework and its related applications such as CICAS-V (traffic signal violation), CICAS-LTA (signalized left-turn assist) and CICAS-TSA (traffic signal adaptation) (*14*). Songchitruksa (*7*) proposed safety performance monitoring using V2I data. However, the study proposed a framework for monitoring safety performance at signalized intersections by means of CV technology. The study only focused on through-vehicle movements because of their relatively well-defined paths and conflict regions. Types of messages commonly used for signalized intersection operations include the Basic Safety Message (BSM) which describe the vehicle kinematics, Map data describing the intersection geometry and SPaT for sending signal status data (*15*).

CV and Speed Advisory at Intersections

Katsaros et al. (*16*) proposed a Green Light Optimized Speed Advisory (GLOSA) application implementation in a typical reference area and presented the performance analysis results using an integrated cooperative ITS simulation platform. The study focused on the improvement of fuel consumption and the reduction of traffic congestion at intersections using wireless communication between vehicle and intersection signals (V2I communication). Vehicles which were equipped with GLOSA were provided with advisory speeds intended to reduce the number of stops at intersections. Stevanovic et al. (*17*) and Tielert et al. (*18*) employed GLOSA in an attempt to minimize delay, fuel consumption, and emissions at intersections by providing advisory speed messages to vehicles to guide them in moving through the green phase. However, even with GLOSA in place, vehicles must come to a stop at the intersection at some point. With the upper and lower limits of speeds at which vehicles can drive, it is impractical for all the vehicles to arrive during the green phase. The present study focuses on providing advisory messages to vehicles that would come to a stop at the intersection due to the end of the green phase.

Methodology

Microscopic Simulation

Researchers and practitioners have used simulation applications widely for various purposes such as comparison of alternatives, analyzing the impact of developments, and cost estimation (*19*). In this study, VISSIM software was used. Through the Component Object Model (COM) interface, the V2I wireless communication between CVs and the signals were modeled using an algorithm implemented through the C2X Application Programming Interface (API).

Study Site

The signalized intersection of Bruce B Downs Boulevard and East Fletcher Avenue (Figure 1), located in Tampa, Florida, and adjacent to the University of South Florida, was selected for the simulation model. The intersection is one of four major signalized intersections located along the Bruce B Downs corridor. The site was chosen because of the availability of naturalistic driving data for different age groups collected by the Strategic Highway Research Program 2 (SHRP2). The naturalistic driving data was used for simulation data input, specifically, speed profiles, and deceleration and acceleration rates.

Data Input

Data employed for this study was obtained from the SHRP2 Naturalistic Driving Study (NDS), through Virginia Polytechnic Institute, for the Bruce B Downs corridor, in Tampa, Florida (*20*). Bruce B Downs corridor is a one-mile stretch containing four major signalized intersections and the highest traversal density of NDS participants, about 400–600 people. This corridor is also one of the several corridors located in Hillsborough County, Florida with a high severe injury crash rate (*21*).

Traffic and signal timing data were provided by the city of Tampa Traffic Office. Geometric information such as number of lanes, lane widths, and turning radii were extracted from Google Earth Pro software. PM peak traffic data were also used for this study. The model was calibrated based on the Florida Department of Transportation (FDOT) simulation guidelines to reflect real conditions (*22*). The volume levels and signal timings are depicted in Table 1.

Simulation Test Bed

The intersection simulation model was developed in VISSIM. V2I communication was modeled using the C2X module available in the software through the COM API. This module enables the modeling of the wireless communication involving high-speed exchange of data

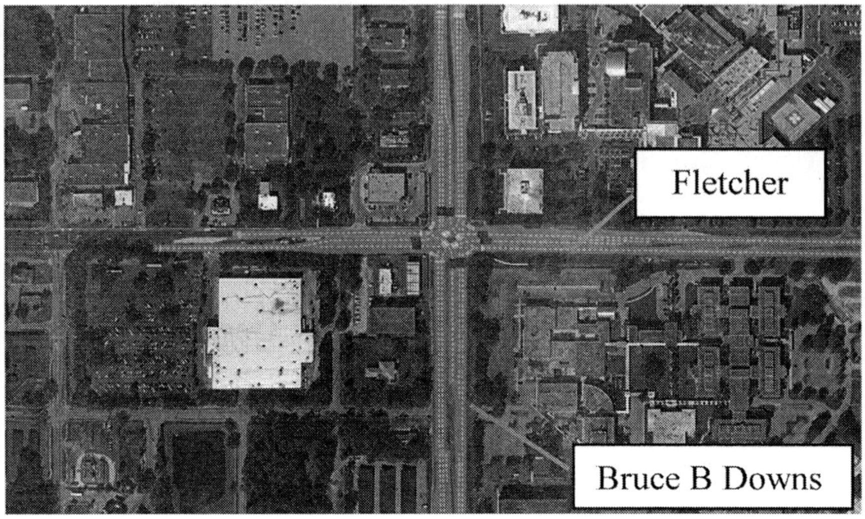

Figure 1. Study site.

Table 1. Traffic Volumes and Signal Timing Settings

Traffic volume												
Phase number	1	2		3	4		5	6		7	8	
Lane type	SBL	NBT	NBR	WBL	EBT	EBR	NBL	SBT	SBR	EBL	WBT	WBR
Volume (veh/hr)	275	1148	243	144	637	204	382	612	320	371	625	254
Signal timing setting												
Min green (sec)	5	5		5	5		5	5		5	5	
Veh extension (sec)	3	3		3	3		3	3		3	3	
Max 1 (sec)	7	19		7	20		7	19		9	18	
Yellow (sec)	3.5	3.5		3.5	3.5		3.5	3.5		3.5	3.5	
Red clearance (sec)	1	1		1	1		1	1		1	1	

plus other wireless communication characteristics such as the wireless transmission success rate. Scripting was done using Visual Basic (VB). The simulation resolution was set to 10 time steps/sim.sec, which is equivalent to the transmission frequency of 10 Hz for the BSM.

The intersection was modeled with pre-timed signal control to simplify the SPaT information tracking. The naturalistic driving data from the SHRP2-NDS was used for calibrating vehicle speeds at the intersection approach.

Several operational assumptions were made to model V2I communications in this study. Some of these assumptions were derived from a study by Songchitruksa (*7*). These assumptions are listed below:

- Although the effective range of V2I communications is over 2000 Ft., communication was maintained for only those vehicles which were in the range of 765 Ft. from the intersection stop bar. This range was established by considering the speed limit of the approach (45 mph) and the Green length + Amber time. A five mph allowance was made for vehicles driving at a speed slightly higher than the speed limit.
- At every time step, the RSE is assumed to process data from all vehicles in the intersection approach as well as the SPaT information. Also the RSE transmits SPaT and Map information to the OBEs where speed advisory messages to each "qualified" C2X equipped vehicle is calculated utilizing an algorithm described in more detail in the next section.
- During the exchange of information with the RSE, the temporary Identification (ID) of the OBEs was assumed to remain the same.
- The study assumed a 100% compliance rate to the speed advisory messages provided.

Table 2. Datasets Produced by the Algorithm

Type of information	Variables collected
SPaT information	• Signal state • Signal state running time • Remaining green time
Map data	• Stop bar location • RSE location
Vehicle data	• Vehicle type • C2X status of C2X equipped vehicles • Vehicle speed • Vehicle acceleration/deceleration • Vehicle position on the intersection approach • Lane of movement

V2I Communication Algorithm

The objective of the algorithm is to provide a smooth deceleration of vehicles that are coming to a stop at the intersection. This is achieved by providing speed advisory messages as soon as the algorithm detects that a vehicle will arrive at the intersection while the signal is red. Thus, the algorithm seeks to resolve the potential for hard-braking as the vehicle approaches the intersection stop bar.

The algorithm was developed only for North-Bound Through (NBT) movements at the study intersection (see Figure 1). Dynamic arrays, populated after every time step, were used for storing the relevant CV data. The algorithm starts by checking if there are vehicles in North-Bound (NB) approach within the range of communication. If there are vehicles approaching the intersection, three types of datasets, shown in Table 2 are produced.

The vehicle position and location of the stop bar were used to establish the vehicle distance to the stop bar. Given d as the distance to the intersection stop bar, u as the current speed of the vehicle, and a as the acceleration of the vehicle, the time to reach the intersection stop bar (T_{TL}) can be calculated using Equation 1:

$$T_{TL} = \begin{cases} \frac{d}{u} & \text{when } a = 0 \\ \frac{-u}{a} + \sqrt{\frac{u^2}{a} + \frac{2d}{a}} & \text{when } a \neq 0 \end{cases} \quad (1)$$

The value of T_{TL} was used to check the signal status by the time the vehicle arrives at the intersection.

If a queue exists at the intersection, drivers would start slowing down if the signal is still green. Therefore, no speed advisory messages would be required. As shown in Figure 2 with accompanying notation, a series of conditions had to be fulfilled before speed advisory messages were given to the driver of a "qualified" C2X vehicle. It is important to note that the advisory messages are provided when the signal is still green. The conditions imposed ensured that speed advisory messages were given to drivers of vehicles that will arrive at the intersection when the signal is red. The definitions of notations used in Figure 2 are shown below.

D	Distance to intersection stop bar
t_{clr}	Time to clear intersection
t_g	Green time
t_{gr}	Remaining green time
t_a	Amber time
V	Current vehicle speed

Stop Distance Estimation

Before the speed advisory messages were provided, the position where a vehicle would come to stop had to be estimated, as shown in Figure 3. This location was then used as a reference for obtaining a suitable deceleration rate to stop the vehicle.

Where: C_i = Vehicle length
H_i = Average standstill distance
D = Vehicle distance to stop bar
d_s = Stopping distance

If there are n vehicles ahead of the current vehicle, the stopping distance is computed as:

$$d_s = D - \sum_{i=1}^{n} C_i - \sum_{i=1}^{n-1} H_i \quad (2)$$

The value of d_s is obtained before the preceding vehicles come to a stop at the stop bar. The values of H were assumed to be fixed at 7.4 Ft which is the average standstill distance of the Wiedemann 74 car following model after calibration.

Braking Deceleration Estimation

When a C2X equipped vehicle is "qualified" for receiving advisory speed messages, a suitable uniform deceleration rate was calculated using the information extracted from the vehicle, plus additional Map data using Equation 3:

$$d_{br} = \frac{V^2}{2 \times d_s} \quad (3)$$

Where:
d_{br} = Uniform deceleration required to bring a vehicle to stop
V = Current speed of the vehicle
d_s = Stopping distance as obtained using Equation 2

The calculated value of d_{br} was then used to calculate the values of speed as the vehicle moves along using

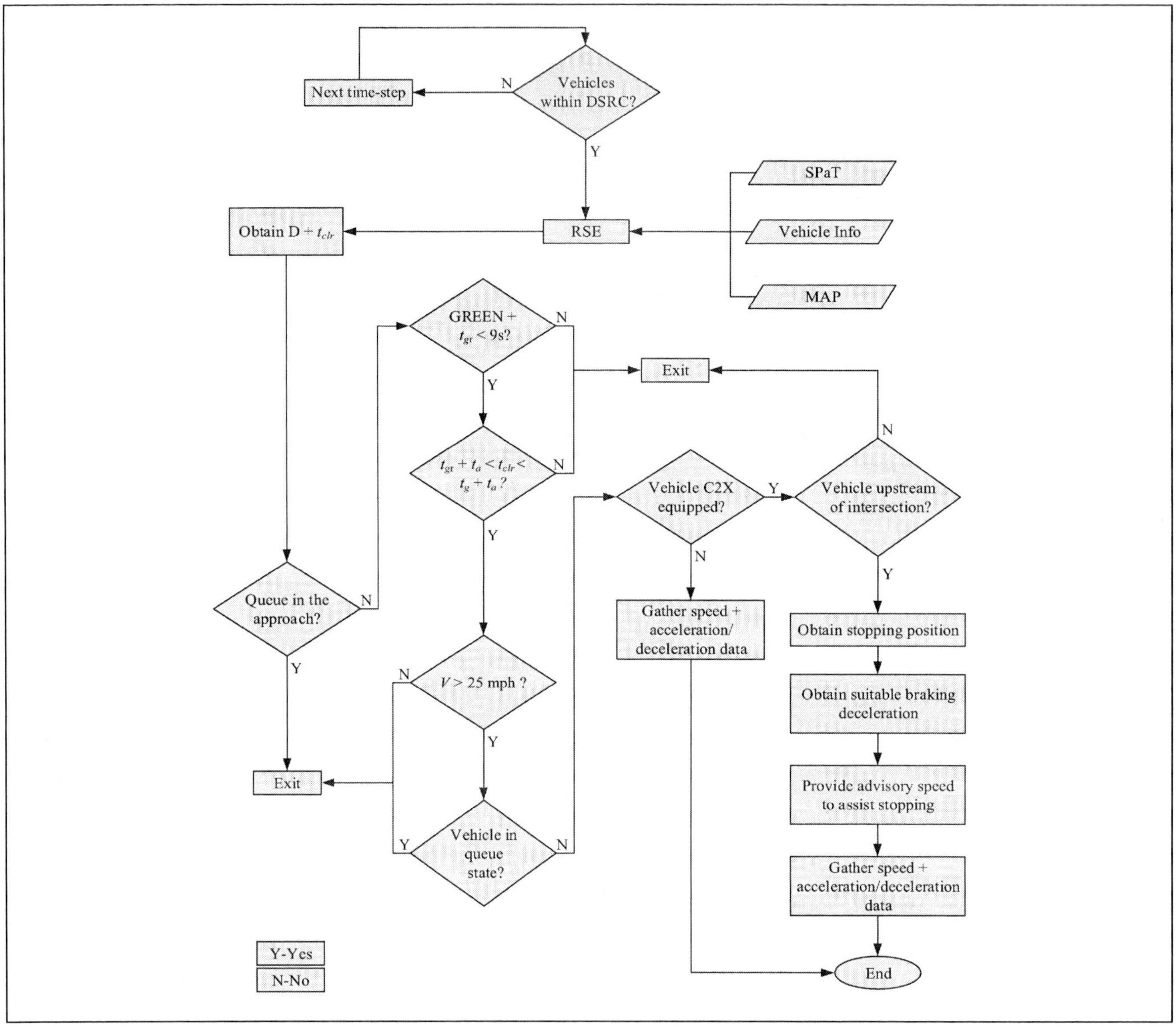

Figure 2. Algorithm for V2I communication.

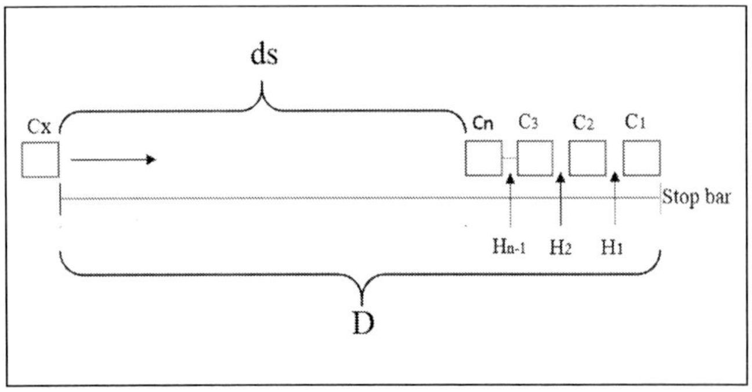

Figure 3. Stopping distance illustration.

Equation 4. These speed values were provided to the driver as speed advisory messages:

$$V_{adv} = \sqrt{2 \times d_{br} \times d_s} \tag{4}$$

Where:

V_{adv} = Advisory speed at the current position

The values of d_s and V_{adv} decrease as the vehicle approaches the intersection. As a result, the speed advisory messages would prompt vehicles to slow down before the signal turns red because the algorithm had already projected that these vehicles would arrive on a red signal indication.

Evaluation

Vehicle Approach-Speed Profiles

The main objective of an ASAS is to enable vehicles that come to a stop at the intersection to smoothly decelerate as a result of receiving advisory messages in advance. Therefore, a record of vehicle speeds as they approach the intersection would be a good measure of how well the objective is met.

Vehicle approach speeds were recorded using detectors modeled in VISSIM. The first detector was positioned at a distance of 750 Ft from the intersection stop bar. This is within the range of V2I communication used in this study (765 Ft). A total of 23 detectors were used to obtain a smooth speed profile at 32.8 Ft (10 m) intervals. Since the study focused on vehicles coming to a stop at the intersection, speeds were recorded only for such vehicles. Speed data for stopping vehicles that were not C2X equipped were also taken into consideration to provide a means for comparison.

Within the communication range, based on vehicle speed and acceleration, different C2X vehicles qualify to receive advisory messages at different locations. Detectors close to the stop bar recorded a higher number of vehicles. To obtain a complete speed profile for decelerating vehicles, a record was kept only for stopping vehicles that were detected from the first detector.

Safety Evaluation

Surrogate safety evaluation is a widely-accepted alternative to historical crash data analysis (*23*). Given the randomness and rarity of crash occurrence, statistics can be applied to relate surrogate safety measures with crash frequency and severity even before crashes occur. Traffic conflicts, a commonly used surrogate safety measure, are defined as situations where two or more vehicles will collide if their movements remain unchanged (*24*). Thus, the number, type, and severity of collisions that occur can be used as an indicator of traffic safety (*25*). This study focused only on rear-end collisions since they are the most prevalent collision type at signalized intersections and can result from hard-braking, a condition addressed here.

Specific thresholds are applied to measurable traffic indicators, such as time to collision (TTC) and post encroachment time (PET), to obtain quantitative data of traffic conflicts. The Surrogate Safety Assessment Model (SSAM), a software application provided by the Federal Highway Administration (FHWA), enables the identification of traffic conflicts by using a statistical analysis of vehicle trajectory files generated from microscopic simulations. Several conflict indicators are provided by the SSAM software based on the trajectory files developed from each scenario run.

For this study, the results from the SSAM did not prove to be useful for several reasons. First, this study specifically focused on the safety evaluation of vehicles that are coming to a stop at the intersection, of which SSAM provided a vast amount of conflict data that was difficult to filter to obtain the required data. Traffic conflicts also occur at different locations, and due to the very specific objective of this study, some attributes were not available in SSAM output to pinpoint conflicts resulting from stopping vehicles. As a result, a separate algorithm which uses a maximum deceleration (MaxD)-based event as a surrogate safety measure had to be developed.

A study by Songchitruksa (*7*) proposed a methodological framework designed to extract and compute safety indicators from vehicle movement, intersection description, and signal data available within the RSE at an isolated signalized intersection. The study classified the safety indicators as either single-OBE or dual-OBE, depending on the number of OBE-equipped vehicles that needed to be monitored. One potential measure of the single-OBE, listed by the study, was the MaxD-based event – the maximum deceleration of a vehicle during an event of continuous braking (*26*). The frequency of these events on a particular roadway section can be related to disturbances in traffic flow, and can indicate the risk of rear-end crash occurrence. According to Songchitruksa (*7*), MaxD-based events are not always a precursor for rear-end crashes. Mixed results of potential indicators can occur with unsafe scenarios. MaxD-based events are triggered by several factors and require careful examination to avoid mixed results. This research work adopted this safety indicator, the MaxD-based event, and refined it to suit the objectives of this study.

VISSIM uses a stochastic, time-step based, microscopic traffic flow model that treats driver-vehicle units as basic entities. It uses the Wiedemann's traffic flow model which is based on the assumption that there are basically four different driving states for a driver: free driving, approaching, following, and braking (*27*). The braking state occurs when a driver applies medium to high deceleration rates if the distance to the preceding vehicle falls below the desired safety distance. This happens if the driver of the preceding vehicle abruptly changes speed or the driver of a third vehicle changes lanes to squeeze between two vehicles in the adjacent lane. Additionally, the built-in driver behavior component in VISSIM has four main states that can be used to determine the driving states of the vehicles in the network. The close-up state describes vehicles closing the distance to a stationary vehicle in front or a hindrance, such as signal heads, stop signs, priority rule, and conflict areas. Since this study focused on a signalized intersection approach, the close-up state could only result

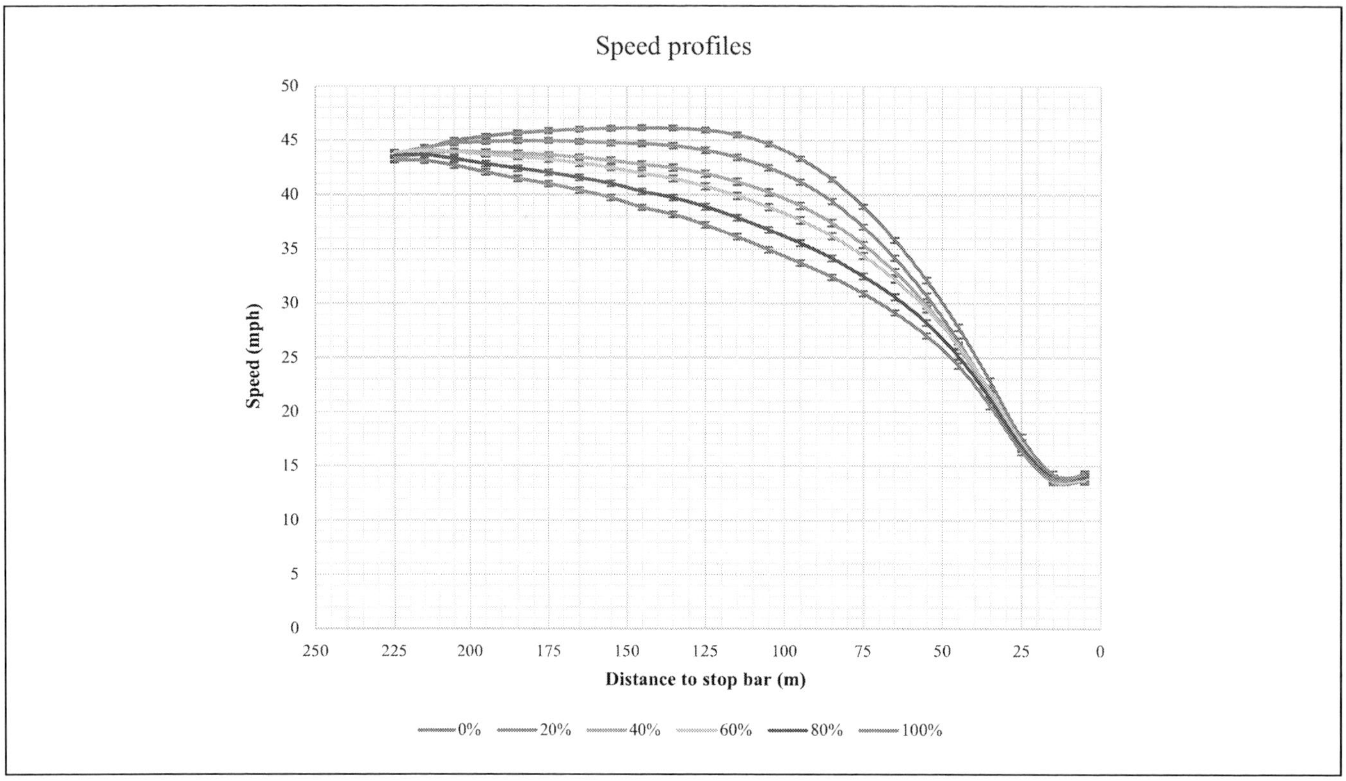

Figure 4. Speed profiles at different penetration rates.

due to a preceding stationary vehicle or a signal head hindrance.

A threshold for recording a MaxD-based event was set to 14.8 Ft/s^2, the emergency deceleration rate in accordance with the AASHTO Green Book (*28*). Vehicles braking harder than 14.8 Ft/s^2 were tracked to the maximum value of deceleration reached, and the value was recorded along with vehicle speed, position, and driving state. The data of hard-braking vehicles, closing up at the intersection approach, were obtained together with the driving state of the vehicles. In the next step, the data were normalized by the total number of vehicles stopped at the intersection.

Results and Discussion

Speed Profiles

Speed profiles, illustrated in Figure 4, were developed using the average speed from each speed detector at a 95% confidence interval. The speed profiles show a clear difference between the base condition (0% penetration or 0% of vehicles with OBE) and when ASAS is implemented at 100% saturation of OBEs. When advisory speed messages were provided to drivers, deceleration started at about 740 Ft (225 m) from the stop bar, nearly 328 Ft (125 m) earlier when compared with the base condition. All of these vehicles will stop at the intersection, but having advance information would help the drivers in smoothening their brake action as they approach the stop bar.

The speed profiles shown in Figure 4 do not include a speed of 0 mph as the detectors placed near the stop bar also recorded speeds of accelerating vehicles during the discharge of the formed queue at the intersection.

Statistical Comparison of Safety Measures

The safety indicator used for statistical comparison was the number of vehicles experiencing hard-braking toward the end of the intersection approach, termed a MaxD-based event. To obtain a valid comparison between the baseline (0% saturation of OBEs) and the 100% saturation of OBEs scenarios, normalization was done by using a selected exposure variable. The exposure variable selected was the number of vehicles within 765 Ft. (communication range of C2X vehicles) of the intersection stop bar during a green signal with no queue formed, but came to a stop upon arrival at the intersection due to a red signal indication. The statistical test procedure suggested by Griffin et al. (*29*) was performed to determine if significant differences between the recorded safety indicators were present. Simulation results and statistical comparison results are summarized in Table 3.

Table 3. Summary of Simulation Results

Measure	Baseline	Scenario 1
Number of MaxD-based conflicts	74	39
Number of stopped vehicles[a]	4460	4450

Statistical comparison of safety indicator

Measure	Baseline	Scenario 1	Z[b]	*p*-value	Change (%)	Statistically significant
MaxD-based conflict rate[c]	16.59	8.76	−3.19	<0.01	−47.2	Yes

[a]Several vehicles within 765 Ft (communication range of C2X vehicles) of the intersection stop bar during a green signal with no queue formed, but came to a stop upon arrival at the intersection due to a red signal indication.

[b]Computed as: $Z = \dfrac{\left(\frac{A+0.5}{EA} - \frac{B-0.5}{EB}\right)}{\sqrt{\frac{A+B}{(E_A+E_B)E_A} + \frac{A+B}{(E_A+E_B)E_B}}}$

Where A = Total count in after period; B = Total count in before period; E_A = Exposure in after period; E_B = Exposure in before period.

[c]Computed as: $\dfrac{\sum Count}{\frac{\sum veh}{1000}}$, in count per 1,000 vehicles. The denominator represents exposure.

The difference in the number of observed MaxD-based conflicts was found to be statistically significant at 95% confidence level. The number of vehicles experiencing hard-braking was reduced by about 50% as a result of the advisory speed messages. This shows that a CV application is a viable solution for reducing the number of rear-end conflicts and possible crashes resulting from the change of signal indication as vehicles come to a stop at the intersection.

Sensitivity Analysis of Market Penetration Rates

Speed Profiles. The effectiveness of the CV deployment depends on the market penetration rate (the overall percentage of vehicles equipped with OBE). Application of ASAS requires vehicles to be equipped with OBE so that they can be detected through V2I communications. Since 100% saturation of OBEs is not expected in the near future, a sensitivity analysis on the effectiveness and reliability of the proposed application was done at different penetration rates in increments of 20% ranging from 0% to 100%. Figure 4 demonstrates how the speed profiles vary with market penetration rates. The profiles indicate that as the number of vehicles that respond to speed advisory messaging increases, the average approach speed of the vehicles decreases.

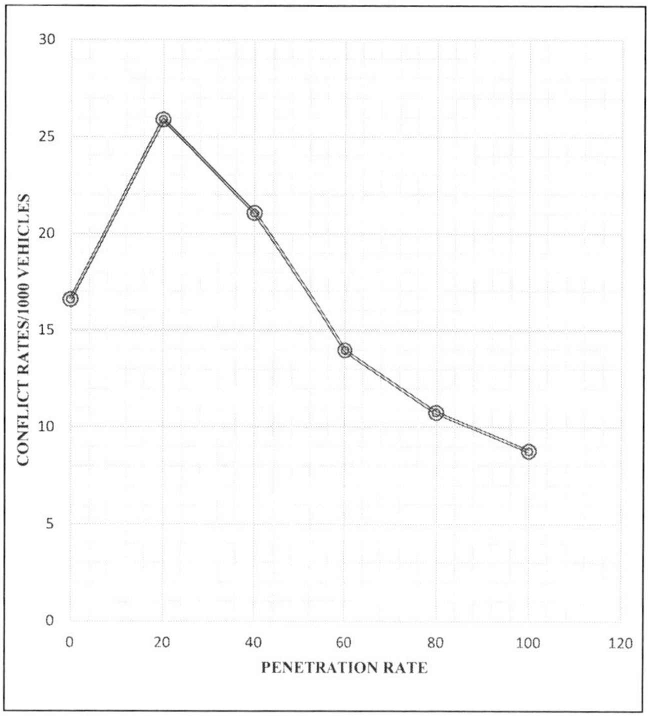

Figure 5. Variation of MaxD-based conflicts with market penetration rate.

Safety Measures. The proposed application was also examined at different saturation rates of the OBEs to obtain a general trend on how the adopted safety indicator varied with varying penetration rates. The results and statistical comparison are shown in Table 4.

The results, as shown in Figure 5, indicate that there is a degradation in the level of safety up to 40% penetration rate when ASAS is implemented at the intersection. The mixture of normal vehicles and C2X-equipped vehicles following the speed advisory messages, which produced more variations in the approach speed of vehicles, is most likely the cause of this degradation. Results indicate that at least 60% saturation of OBEs is required to observe tangible improvements.

Table 4. Summary of Simulation Results at Different Penetration Rates

Penetration rate	Number of MaxD-based conflicts	Number of stopped vehicles
Baseline	74	4,460
20%	115	4,441
40%	94	4,462
60%	62	4,438
80%	48	4,459
100%	39	4,450

Statistical comparison of safety indicator at different penetration rates

Penetration rate	MaxD-based conflict rate[a]	Z[b]	p-value	Change (%)[c]	Statistically significant at 95% CL
0%	16.59				
20%	25.89	3.08	<0.01	56.1	Yes
40%	21.07	1.62	0.105	27.0	No
60%	13.97	–0.91	0.362	–15.8	No
80%	10.76	–2.26	0.024	–35.1	Yes
100%	8.76	–3.19	<0.01	–47.2	Yes

[a]Computed as in Table 3.
[b]Computed as in Table 3 with reference to the base condition.
[c]Computed with reference to the base condition.

Conclusions and Recommendations

This paper proposed a CV application which utilizes V2I communications at signalized intersections. The application aims at reducing the risk of rear-end crashes at signalized intersections caused by the indecision of drivers when a signal indication changes. The proposed application makes use of the information received at the RSE from vehicles' OBE through V2I communications in the form of BSM, as well as Map and SPaT information. An algorithm was developed to provide advance speed advisory messages to C2X-equipped vehicles projected to arrive at the intersection on a red signal indication.

A microscopic simulation approach was used for evaluating the effectiveness of the proposed system. VISSIM and the built-in C2X module were used in the analysis. The script was written in VB to implement V2I communications. First, two simulation scenarios were designed, one without the application of ASAS, and one with the application of ASAS at 100% market penetration rate (saturation of OBEs). Speeds of vehicles approaching the intersection, as well as the number of vehicles closing in that experienced hard-braking toward the end of the intersection approach, were recorded. This provided data used to evaluate the effectiveness of the proposed application. The evaluation results indicate that ASAS has a potential of reducing the number of MaxD-based conflicts considerably. The results also revealed a reduction of about 50% in number of MaxD-based conflicts resulting from vehicles receiving advance speed advisory messages.

Since full market penetration of OBEs is not expected soon, a sensitivity analysis of market penetration rates was performed and showed a degradation in safety conditions at low penetration rates. The results demonstrated that at least 60% saturation of OBEs is required to observe a reduction in the number of MaxD-based conflicts.

In future work, more research is needed to investigate the response of drivers to the speed advisory messages received. This study assumed 100% compliance rate of the received messages. The algorithm could also be modified to work with actuated signal control as this study only dealt with pre-timed signals. The V2V communication and a mechanism of communicating messages received by C2X-equipped vehicles to non-equipped vehicles visually at higher and lower market penetration rates respectively can be researched to potentially observe safety improvements.

Author Contributions

The authors confirm contribution to the paper as follows: study conception and design GN, TS, SS, EM, EEO, JS; data collection: GN, TS, SS; analysis and interpretation of results; GN, EEO, JS, EM; draft manuscript preparation: GN, TS. All authors reviewed the results and approved the final version of the manuscript.

References

1. AASHTO. *Background Information on Proposed Policy Resolution AASHTO SPaT Challenge*. AASHTO, Washington, D.C., 2016.
2. Roess, R. P., E. S. Prassas, and W. R. McShane. *Traffic Engineering*. Pearson Higher Education, Upper Saddle River, NJ, 2011.
3. Nuyttens, N., A. Carpentier, K. Declerq, and E. Hermans. *Jaarrapport Verkeersveiligheid 2012: Analyse van*

Verkeersveiligheidsindicatoren in Vlaanderen Tot En Met 2012. Policy Research Centre for Traffic Safety & Belgian Road Safety Institute, 2014. http://www.steunpuntverkeersveiligheid.be/nl/node/574.

4. Traffic Safety Facts. *A Compilation of Motor Vehicle Crash Data from the Fatality Analysis Reporting System and the General Estimate System*. National Highway Traffic Safety Administration, Washington, D.C., 2013.
5. Aarian Marshall. *Enlighten App Uses AI to Predict When Lights Will Turn Green*. https://www.wired.com/2016/10/enlighten-app-uses-ai-predict-lights-will-turn-green/. Accessed May 11, 2017.
6. *Siemens to Provide Vehicle-to-Infrastructure Technology for Tampa-Hillsborough Expressway Authority Connected Vehicle Pilot Deployment*. Siemens USA Newsroom. http://news.usa.siemens.biz/press-release/mobility/siemens-provide-vehicle-infrastructure-technology-tampa-hillsborough-expressw. Accessed May 11, 2017.
7. Songchitruksa, P., and L. Zha. Advancing Safety Performance Monitoring at Signalized Intersections through Use of Connected Vehicle Technology. *Transportation Research Record: Journal of the Transportation Research Board*, 2014. 2432: 99–109.
8. SAE. *Dedicated Short Range Communications (DSRC) Message Set Dictionary*. 2013, pp. 724–776. https://doi.org/10.4271/J2735_201603.
9. Kenney, J. B. Dedicated Short-Range Communications (DSRC) Standards in the United States. *Proceedings of the IEEE*, Vol. 99, No. 7, 2011, pp. 1162–1182. https://doi.org/10.1109/JPROC.2011.2132790.
10. *Connected Vehicle*. FHWA Traveler Information. https://ops.fhwa.dot.gov/travelinfo/infostructure/aboutinfo.htm. Accessed July 22, 2017.
11. Goodall, N., B. Smith, and B. Park. Traffic Signal Control with Connected Vehicles. *Transportation Research Record: Journal of the Transportation Research Board*, 2013. 2381: 65–72.
12. Ilgin Guler, S., M. Menendez, and L. Meier. Using Connected Vehicle Technology to Improve the Efficiency of Intersections. *Transportation Research Part C: Emerging Technologies*, Vol. 46, 2014, pp. 121–131. https://doi.org/10.1016/j.trc.2014.05.008.
13. Wassim, N., J. Koopmann, J. D. Smith, and J. Brewer. *Frequency of Target Crashes for IntelliDrive Safety Systems*. National Highway Traffic Safety Administration, Washington, D.C., 2010, p. 50. https://www.nhtsa.gov/DOT/NHTSA/NVS/Crash%20Avoidance/Technical%20Publications/2010/811381.pdf.
14. Misener, J. A. *Cooperative Intersection Collision Avoidance System (CICAS): Signalized Left Turn Assist and Traffic Signal Adaptation*. California PATH Research Report. University of California at Berkeley, Berkeley, CA, 2010.
15. Zha, L., Y. Zhang, P. Songchitruksa, and D. R. Middleton. An Integrated Dilemma Zone Protection System using Connected Vehicle Technology. *IEEE Transactions on Intelligent Transportation Systems*, Vol. 17, No. 6, 2016, pp. 1714–1723. https://doi.org/10.1109/TITS.2015.2490222.
16. Katsaros, K., R. Kernchen, M. Dianati, and D. Rieck. Performance Study of a Green Light Optimized Speed Advisory (GLOSA) Application using an Integrated Cooperative ITS Simulation Platform. *Proc., 2011 7th International Wireless Communications and Mobile Computing Conference*, Istanbul, Turkey, 2011, pp. 918–923.
17. Stevanovic, A., J. Stevanovic, and C. Kergaye. Impact of Signal Phasing Information Accuracy on Green Light Optimized Speed Advisory Systems. *Transportation Research Record: Journal of the Transportation Research Board*, 2013. 2390: 53–59.
18. Tielert, T., M. Killat, H. Hartenstein, R. Luz, S. Hausberger, and T. Benz. The Impact of Traffic-Light-to-Vehicle Communication on Fuel Consumption and Emissions. *Proc., 2010 Internet of Things (IOT)*, Tokyo, Japan, IEEE, New York, NY, 2010, pp. 1–8. https://doi.org/10.1109/IOT.2010.5678454.
19. Lownes, N., and R. Machemehl. Sensitivity of Simulated Capacity to Modification of VISSIM Driver Behavior Parameters. *Transportation Research Record: Journal of the Transportation Research Board*, 2006. 1988: 102–110.
20. Virginia Tech Transporatation Institute. *SHRP2 NDS Data Access*. https://insight.shrp2nds.us. Accessed June 22, 2016.
21. Tindale-Oliver and Associates Incorporation. City of Tampa Mobility Plan. Tampa, 2012 https://www.tampagov.net/sites/default/files/transportation/files/MOBILITY_PLAN.PDF. Accessed July 10, 2016.
22. Florida Department of Transportation. *Traffic Analysis Handbook: A Reference for Planning and Operation*. Tallahassee, FL, 2014.
23. *Transportation Research Circular E-C179: Theory, Explanation, and Prediction in Road Safety: Promising Directions*. Transportation Research Board of the National Academies, Washington, D.C., 2013. http://onlinepubs.trb.org/Onlinepubs/circulars/ec179.pdf.
24. Amundsen, F., and C. Hyden. Proc., The First Workshop on Traffic Conficts. Norwegian Council for Scientific and Industrial Research, Oslo, Norway, 1977.
25. Wang, C., and N. Stamatiadis. Surrogate Safety Measure for Simulation-Based Conflict Study. *Transportation Research Record: Journal of the Transportation Research Board*, 2013. 2386: 72–80.
26. Allen, B. L., B. T. Shin, and P. J. Cooper. Analysis of Traffic Conflicts and Collisions. *Transportation Research Record: Journal of the Transportation Research Board*, 1978. 667: 67–74.
27. PTV AG. *VISSIM 8 User Manual*. PTV AG, Karlsruhe, Germany, 2016.
28. AASHTO. *Policy on Geometric Design of Highways and Streets*. AASHTO, Washington, D.C., 2001.
29. Griffin, L. I., and R. J. Flowers. *A Discussion of Six Procedures for Evaluating Highway Safety Projects*. FHWA, U.S. Department of Transportation, Washington, D.C., 1997.

The Standing Committee on Vehicle-Highway Automation (AHB30) peer-reviewed this paper (18-05570).

Article

TRR
JOURNAL OF THE TRANSPORTATION RESEARCH BOARD

Pedestrians/Bicyclists and Autonomous Vehicles: How Will They Communicate?

Transportation Research Record
2018, Vol. 2672(22) 58–66

DOI: 10.1177/0361198118777091
journals.sagepub.com/home/trr
SAGE

Sergiu C. Stanciu[1], David W. Eby[1], Lisa J. Molnar[1], Renée M. St. Louis[1], Nicole Zanier[1], and Lidia P. Kostyniuk[1]

Abstract

Interpersonal roadway communication is a vital component of the transportation system. Road users communicate to coordinate movement and increase roadway safety. Future autonomous vehicle research needs to account for the role of interpersonal roadway communication. This literature review synthesizes research on interpersonal interaction between drivers, bicyclists, and pedestrians while also directing attention to implications for autonomous and connected vehicle research. Articles were collected from TRID, PsycINFO, Google Scholar, and ScienceDirect using search terms relevant to driving, communication, and vulnerable road users. The synthesis documents that interpersonal communication not only takes place but is also an important and understudied aspect of safe roadway travel. The review also found that road users employ a variety of communication methods that include gestures, facial expressions, and built-in vehicular devices. Comprehension of messages is influenced by several factors including culture, context, and experience. These results shed light on potential issues and challenges of interpersonal communication and the introduction of autonomous vehicles to the roadway.

Interpersonal communication between road users is an integral aspect of using the transportation system. Whether driving a motor vehicle, riding a bicycle, or crossing the street, safe transportation requires that everyone share the road. Communication between road users helps to coordinate behaviors, regulate road use, signal intent, and encourage drivers to behave in certain ways (*1*, *2*). These interactions can involve, reminding road users of formal traffic rules when they appear to be breaking them, or using eye contact to establish mutual acknowledgment of what is occurring, such as when a pedestrian crosses the street and a driver yields to him or her. Zaidel (*3*) argues that the road can be considered a social environment, where driver decision-making is partly determined by the way that drivers communicate with each other. Road users consider such communication important, and when drivers fail to communicate their intent, other road users view their behavior less favorably (*4*).

In response to the critical need for interpersonal communication, modern automobiles are equipped with technologies to facilitate communication between drivers, such as turn signals, hazard lights, and car horns. For example, drivers are required to indicate their intent to turn or change lanes through the use of a turn signal to other road users, yet research shows that turn signals are not consistently used. Sullivan, Bao, Goudy, and Konet (*5*) conducted a naturalistic investigation of turn signal use at intersections and found that signals were only used 71–75% of the time. When they were used, turn signals alone were not sufficient to judge a driver's intersection behaviors because signal use was partly determined by the road type, turn direction, and state of the surrounding traffic. In recognition of the need for interpersonal communication in the driving environment, other technologies have been developed to further extend communication options while driving, such as remote-controlled expressive car signs that utilize mountable screens that can display short messages or express emotions to other drivers (*6*).

The development of automated and connected vehicles comes with promise that driver-error-related crashes will be reduced. The National Motor Vehicle Crash Causation Survey (NMVCCS) found that about 94% of crashes were due to driver error (*7*), and advocates of automated vehicles have predicted that automated vehicles will reduce number of crashes by 80% (*8*). However, even with just a few autonomous vehicles on the road at this time, an automated vehicle was already involved in a crash recently when it attempted to merge into traffic, to maneuver around an obstacle on the road, as a bus was approaching from behind (*9*). The autonomous vehicle's programming assumed that the bus driver would yield when the autonomous vehicle attempted to merge into traffic; yet, this assumption was incorrect, and the

[1]University of Michigan Transportation Research Institute, Ann Arbor, MI

Corresponding Author:
Address correspondence to Sergiu C. Stanciu: sergiucs@umich.edu

conflict could not be resolved through interpersonal communication.

At present, automated vehicles do not have the same capacity as drivers to communicate with other road users. For example, automated vehicles do not have a means to signal intent beyond the use of a turn signal, they cannot indicate whether or not they intend to yield, and they cannot yet read and interpret gestures from vulnerable road users (*10*). Considering how ubiquitous interpersonal interaction is within the context of driving, this is a potential safety concern worth addressing.

Interpersonal communication has to do with the exchange of information between people that enhances mutual understanding of a situation, indicates behavioral intent, or responds to the actions of another. Effective communication involves all parties communicating their intent in a manner that is received and understood. Alternatively, miscommunication is either a failure to communicate intent effectively (e.g. clearly visible, easily understood) or a misunderstanding of a communication attempt. Lack of communication is the failure to communicate intent or response within a meaningful timeframe.

Lack of communication and miscommunication on the road also represent safety concerns. The NMVCCS found that "false assumption of others' actions" was the critical reason for 5% of all crashes, of which at least a part could be likely attributed to miscommunication between road users while driving (*7*). The safety effects of miscommunication may be even greater when vulnerable road users are considered. In 2014, 26% of pedestrian fatal crashes and 27% of bicycle fatal crashes, were attributed to "failure to yield right of way", with a large proportion of these crashes likely attributable to lack of communication or miscommunication (*11*). Despite these statistics, there is not a good way to estimate the prevalence of crashes for which communication problems are primary causal factors, and, thus, a full grasp of the safety aspects of miscommunication or the lack of communication in the transportation system is lacking.

Currently, our transportation system is not only undergoing a technological transformation but also a social-communicative transformation. As the development and deployment of automated and connected vehicles continues, there will be a mixed fleet of autonomous, semi-autonomous, and non-automated vehicles, as well as vulnerable users (e.g., pedestrians and bicyclists) that all must safely interact with each other. Thus, it is critical to better understand the nature of how, why, and when people communicate with each other on the roadway. Such information will be useful for developing future vehicles and behavioral countermeasures that will not only prevent crashes but also improve the well-being of all road users.

Objectives of This Paper

The objective of this research is to further the understanding of the role of communication among road users through a scoping synthesis of literature on this topic. Because of the general lack of randomized control trials or the use of comparison groups in this field, we deemed that a meta-analysis of the literature would not yield meaningful results. The intent is to identify issues and challenges within the context of interpersonal communication on the road that developers of autonomous and connected vehicles need to consider and address in their development of vehicles that will interact safely with vulnerable road users.

This paper is organized as follows: the next section describes the methods used in the synthesis of the literature. This is followed by the results, organized into the following three themes: types of interpersonal communications, communications between drivers and vulnerable road users, and comprehension of interpersonal messages. The last section is a summary of the findings and a discussion of the implications of these findings for autonomous vehicles.

Methods

Our search consisted of three steps. In the first step, we defined the forms of interpersonal communication on the roadway to be included in our synthesis, based on relevant literature (*1*, *4*) and the research team's background. The search terms used to find research materials were related to driving (driver, vehicle, driving), communication (gestures, interpersonal communication, interaction, communication), and type of communication (turn signals, horn honking, hand gestures, eye contact). The search terms were chosen based on relatedness to commonly used methods of interpersonal communication and general driving terms. Next, articles were searched in TRID, PsycINFO, Google Scholar, and ScienceDirect databases. From these databases, relevant journal articles, technical reports, and conference papers were gathered. An Internet search was also conducted to find news articles and web pages pertaining to the topic. While there was no year restriction on the search, most articles were published after the year 2000. Finally, manual searches of the reference lists of relevant articles were conducted to collect additional articles that appeared to be appropriate. Overall, the search yielded 66 candidate articles.

These 66 articles were examined by the research team. To be eligible for further review, articles needed to: 1) provide findings related to the types, meanings, and use of various forms of on-road communication, and 2) focus on person–person or vehicle–person communication, excluding vehicle–vehicle or vehicle–infrastructure communication. This process yielded 39 articles for inclusion. The research described in the 39 articles utilized a wide range of data collection techniques including focus groups, questionnaires, naturalistic driving, and driving simulation. The final articles were categorized into several topics relevant to understanding the function of interpersonal communication on the road. These topics were: the ways in which people communicate; communication with vulnerable road users; how well people understand roadway communication and how communication influences driver behavior.

Results

Types of Interpersonal Communication

There are a variety of ways that road users can communicate with each other while sharing the road. Renge (*1*) divided roadway communication into three categories: formal device-based signals; informal device-based signals; and informal gesture-based signals. Renge et al. (*12*) added a fourth category—formal gesture-based signals, which encompasses the arm gestures that bicyclists can use to indicate the direction they intend to turn. Formal device-based signals include the use of hazard lights, turn signals, and car horns for their intended use, while informal signals can vary widely, including the use of hand gestures, eye contact, or vehicle headlights. Even with formal device-based signals, the meaning can change depending on the context in which a signal is used, as is the case with the car horn. Use of honking can generally serve five different functions: as social etiquette, giving emphasis, as an expression of displeasure, giving an order, or giving notice to something (as cited in *13*). For example, Takada, Fukuda, and Iwamiya (*14*) found that light tapping of the car horn was a means for Japanese drivers to show gratitude for right-of-way yielding. Given the wide range of uses for the car horn, it is important to keep in mind that the situational context is an important component to interpreting horn honking.

Turn signal use also depends on situational context. Faw (*15*) conducted two observational studies of turn signal use rates. The first study was conducted at 22 intersections over the course of 4 months and the second at 2 intersections over 28 months. Turn signal use rates were measured under a variety of conditions (e.g., region, traffic volume, direction of turn). Faw (*15*) found that turn signal use rates were higher in moderate traffic (92%) and lower in light traffic (88.7%) and heavy traffic (78.9%). Drivers also signaled left turns (89.2%) more often than right turns (85.1%). The study also found that drivers signaled 7% less often when closely following another driver, although when drivers were following a vehicle that had a signal on, they activated their signal 6% more often. Sullivan et al. (*5*) obtained similar results after analysis of extensive naturalistic driving data for 108 drivers. These researchers found that drivers were 1.4 times more likely to use their turn signal when turning left as compared with right, which they suggested could be due to greater perceived risk with left turns. They also found that people were about 5 times more likely to use their turn signal on major and minor surface roads compared to local roads, which tend to have less traffic. In addition, the study found that turn signal use at an intersection was 1.5 times more likely if drivers were behind a lead vehicle at the intersection. Thus, there is good evidence that signal use varies by several factors and Sullivan et al. (*5*) concluded that signal use alone may not be a reliable indicator of a persons' intended action. In other words, a lack of signal use does not necessarily mean that the driver does not intend to make a turn.

Unlike formal device-based signals, informal signals do not have any official standard, they are learned through observation and experience (*1*). Hand gestures are a common category of informal communication. For example, drivers may use hand gestures to encourage other road users to go or to halt (*16*), while pedestrians may also use hand gestures to encourage other drivers to yield and to show gratitude for yielding behavior (*16–18*). Because different gestures can be used to convey the same message, research has been conducted to discover what kinds of gestures communicate behavioral intent more effectively (*17*, *18*). Turnbaugh and Turnbaugh (*19*) suggested that road users use hand signals as an internalized form of reciprocal acknowledgment or as a physical expression of monitoring one another. Eye contact, though subtle, is another powerful form of informal communication. Eye contact is processed as a social cue; that is, a demonstrative signal that communicates who is being addressed, as well as that the ensuing action or information will be meaningful (*20*). Šucha (*16*) confirmed that pedestrians and drivers both make use of eye contact to communicate and this can be used by pedestrians to encourage other drivers to yield (*21*).

Informal device-based signals are another potential mode of communication. For example, some drivers use headlights to communicate with other road users (*16*). Renge (*1*) outlined several examples of headlight flashing for communication with other road users including: blinking of headlights to yield one's way, blinking of headlights to another car that is cutting in, and blinking of headlights for thanks. Formal device-based signals, formal gesture-based signals, informal device-based signals, and informal gesture-based signals are four broad categories that can be used to classify the diverse range of methods that road users can use to communicate. Being able to detect and classify the use of signals to communicate is the first step in developing a deeper understanding of interpersonal roadway communication. However, as with other modes of interpersonal communication, the context in which the signaling takes place often dictates the communication message. Depending on the context, the flashing of one's headlights can have different meanings, such as indicating there are police or obstacles ahead, urging a driver to speed up, signaling that it is okay to merge, and numerous other messages. Therefore, the context of messaging plays a critical role in effective interpersonal communication on the roadway.

Communication between Drivers and Vulnerable Road Users

Vulnerable road users also have methods they use to communicate their presence and behavioral intent to drivers. Most vulnerable road users do not have device-based means of communicating intent to drivers and so they often employ informal means of communication. Research cataloguing the use and effects of communication from vulnerable road users has focused on pedestrian street crossing and bicyclist–vehicle interaction.

A common area for pedestrian safety is crossing a road when there is traffic present and there is no signalized crossing. Research shows that drivers are unlikely to yield to pedestrians (see e.g. *22–26*). For example, a study in Malaysia observed pedestrian–driver interactions at a marked crosswalk utilizing cameras focused on the intersection (*25*). Video analysis showed drivers of automobiles yielded to pedestrians in only 1.5% of attempted crossing and motorcyclists never yielded to pedestrians. Another study conducted in Boston found that at unsignalized, marked crosswalks on roads with 40 mph speed limits the average driver yielding rate was only 15% (*22*).

One way in which pedestrians can improve driver yielding is through communication with drivers at points of roadway crossing, such as crosswalks. Pedestrians can communicate with drivers by using their arms, making eye contact, and smiling; drivers can respond by using hand gestures, flashing their lights, or making eye contact (*16*).

The importance of interpersonal communication in driver yielding to pedestrians has been demonstrated in a study by Lehsing, Benz, and Bengler (*27*). They conducted a study in which a driving simulator was linked to a computer that allowed the experimenter to control the actions of a pedestrian (called an avatar in the study) that was visible to the study participant as they drove the simulator. The experimenter controlled the pace and walking direction of the pedestrian (e.g., turning to face the driver) in response to the participant driver's actions, thereby allowing for a form of interpersonal communication to take place. In other conditions, the pedestrian (called a bot in the study) was simply programmed to move along a sidewalk and occasionally cross the street. The study measured driving behavior around the two different types of pedestrians. The researchers found that drivers drove more cautiously in the presence of the bot-pedestrian walking on the sidewalk when compared to the avatar-pedestrian, and at marked crosswalks drivers slowed significantly more for the avatar-pedestrian. The researchers suggested that drivers were more cautious around the bot-pedestrian because drivers lacked interpersonal communication with the bot and therefore did not have information about what the bot might do. On the other hand, participants noticed that avatar-pedestrians displayed safety-oriented behavior that was related to the participant's driving behaviors and this communication allowed participants to better predict the avatar-pedestrian's behavior both while walking on the sidewalk and crossing at marked locations. These results demonstrate the importance of interpersonal communication between drivers and pedestrians.

Research conducted on the effects of signals that pedestrians use to communicate with drivers sheds further light on the nature of this interaction. Studies show that gestures are a common means of informal communication that improves driver-yielding behavior for pedestrians (*17*, *18*). Zhuang and Wu (*18*) conducted a study measuring the difference in yielding rates among several gestures and concluded that visibility and clarity of the gesture were the most important aspects of an effective gesture and that many gestures were misinterpreted. Only one gesture, a left-arm, bent elbow with the palm facing the oncoming driver, was found to significantly increase yielding (12.9%). In a similar study, Crowley-Koch, Van Houten, and Lim (*17*) found that a raised arm "halt" gesture increased driver yielding at crosswalks by 29% compared to a no-gesture condition. They reasoned that this was because the raised arm was an assertive gesture, and that pedestrian assertiveness played a role in the likelihood of driver yielding. Additionally, these researchers suggested that the raised arm gesture could be more easily transitioned to a "thank you" wave, which could encourage future yielding behavior.

There is some evidence that an expression of gratitude might increase driver yielding. Nasar (*28*) conducted a naturalistic study in which drivers passing through two intersections had to choose whether to yield to pedestrians who were confederates for the study. If the driver yielded to the confederate at the first crossing, then they held up a written sign indicating gratitude and they smiled at the driver. If the driver did not yield, at the next crosswalk another confederate presented a sign indicating disapproval and a plea to yield in the future. This treatment was conducted over the course of 3 weeks, and for each of the first 2 weeks there was an 8% and 7% increase, respectively, in yielding at the second crosswalk, with a 5% average increase in yielding at the first crosswalk by the third week. While a modest increase, the results suggest that "thank you" hand waves or other gestures of gratitude might increase yielding behavior of drivers over the long-term.

Other forms of communication that pedestrians can employ to indicate intent to cross the road and have the driver yield include direct gaze and facial expression. Direct gaze can function as both an emotional and social cue, and therefore can be used by pedestrians to get the attention of drivers and communicate their intentions (*20*). In another study utilizing confederate pedestrians at crosswalks, researchers found that when pedestrians stared directly at approaching drivers, drivers yielded 12.6% more often as compared with conditions when they did not stare, and the effect was greater when the confederate was female (*21*). These same researchers (*29*) investigated the effect of a pedestrian's smile on yielding behavior, again using confederates as pedestrians. They found that smiling increased yielding behaviors at marked crosswalks by 12.8% as compared with a neutral expression, but drivers stopped 13.5% more often when the confederate was female. Further, male drivers stopped 20.6% more often for female confederates. Finally, smiling also increased yielding behavior by 11.2% at unmarked crosswalks. These studies show the importance of nonverbal social communication for pedestrians and the influence of gender in pedestrian yielding behavior.

Drivers also communicate with pedestrians to indicate their intent. Šucha (*16*) conducted focus groups with drivers to identify communication strategies used with pedestrians intending to cross the road. Drivers indicated that they use hand waves, light flashes, and eye contact to communicate with pedestrians. Two methods that drivers used to indicate that they did not intend to yield for pedestrians involved the use of relative speed: maintaining their speed or accelerating. Schneemann and Gohl (*30*) found that pedestrians were more likely to feel safe crossing the road if drivers visibly decelerated. Drivers were less likely to yield for the pedestrian if the pedestrian did not react to drivers' deceleration, which is used to show intent to yield. Drivers interviewed by Šucha (*16*) noted several conditions under which they would not yield for a pedestrian, a few of which dealt with interpersonal communication: if the pedestrian notices the driver and if the pedestrian does not cross but waits for the driver to pass. Additionally, Schneemann and Gohl (*30*) found that drivers were more likely to yield when a pedestrian made eye contact with approaching drivers and moved forward because it showed intent to cross.

Since pedestrians and autonomous vehicles do not have access to many of the previously described means to communicate, some research has been conducted to determine how pedestrians might interpret signals given from autonomous vehicles. Clamann, Aubert, and Cummings (*31*) had participants cross the street while a presumed autonomous vehicle approached. The autonomous vehicle had a mounted display that showed one of three symbols indicating to pedestrians whether to cross or not, or the vehicle's current speed. Pedestrian crossing behavior was tested across multiple conditions: approaching vehicle speed, distance from vehicle, crosswalk vs. midblock crossing, and display mode. The researchers found that only 12% of participants had their decision to cross the road influenced by the mounted display, but 46% of participants stated that they would find similar displays useful on autonomous vehicles. The factor that determined pedestrian crossing time the most was the distance gap between pedestrians and the approaching vehicle, with much variance attributable to individual differences. These results indicate that communication of intent may help road users become more comfortable around autonomous vehicles. Additionally, though these results may seem to suggest that communication is not a relevant factor in pedestrian safety, they do highlight the importance of scrutinizing proposed solutions for autonomous vehicle communication, because they may not work as intended.

Bicyclists also need to communicate with drivers to use the road safely. Since bicyclists interact with drivers at crosswalks and on the road, it is important that bicyclists not only make drivers aware of their presence but also communicate behavioral intent, such as when they are about to turn. Studies show that gaze cues can make drivers more aware of bicyclists and their intentions because drivers tend to fixate on a bicyclist's face before directing their attention to gestures or other nonverbal signals (*32*, *33*). Walker (*32*) conducted a laboratory experiment in which drivers had to react to bicyclists at intersections by judging their intentions and reacting appropriately. In some cases, bicyclists gave no signal indicating their intent, in other cases they gazed at the driver or used arm signals. When bicyclists used clear arm signals, drivers were far less likely to stop unnecessarily, indicating that they properly understood the bicyclist's intent. Walker (*32*), however, found a potential problem with signaling: clear arm signals were the best perceived and understood but also required the most cognitive processing, which slowed down reaction times and led to more collisions. Thus, there was a trade-off between the complexity of the mode of communication and the reaction time required to respond appropriately. This finding does not mean that giving clear communication of intent is counterproductive, because it does improve yielding rates and enhances shared understanding of intended behaviors, but it also means this communication strategy should be started well in advance of the intended behavior.

Comprehension of Interpersonal Message

Given the variety of ways in which road users attempt to communicate with each other, comprehension is a critical component to this interpersonal communication. Archer (*34*) demonstrated through video recordings of gestures from around the world that nonverbal communication varies widely between cultures and even identical gestures can have different meanings in different cultures. Factor, Mahalel, and Yair (*35*) presented a "social accident" model to study the effects of social and cultural factors on crashes. They argued that different social groups have different cultural characteristics, which can change the way that driving situations are interpreted. Renge et al. (*12*) conducted a cross-cultural study in Germany, Finland, and Japan to examine differences in comprehension for formal and informal communication signals. Participants were presented with 22 videotaped scenes of road users employing different forms of communication and had to determine the correct meaning behind the signal use, rate their confidence in their response, and rate the presented signals on different scales of friendliness/aggression and necessity. As compared with Finnish and German drivers, Japanese drivers identified the correct meanings of formal signals less often and had lower confidence scores for their answers. Judgements of informal signals revealed several cross-cultural differences in comprehension. For example, in one scenario a motorcyclist was forced to stop when they were abruptly cut off by a driver of an automobile. The motorcyclist then protested by raising his middle finger. The German and Finnish drivers generally understood this signal as a message to obey the traffic regulations and as a sign of aggression. Japanese drivers, on the other hand, did not understand the gesture, with the majority interpreting the hand gesture as "thank you." Many Japanese drivers also did not understand signals involving hazard lights and bicyclists' turn signals. Finnish drivers were the most confident in their comprehension of communication, followed by German and then Japanese drivers. Differences in the comprehension of interpersonal messages can also occur within a country. For example, studies have shown that that horn honking conveys different meanings in different places in the United States (see e.g., *36*, *37*).

Currently and for the foreseeable future, roadway transportation can be considered a social environment, where behaviors are influenced by road users, formal traffic rules, and social norms. Therefore, both the immediate situation and past experiences factor into how roadway users may interpret interpersonal communication. Renge (*1*) found that as compared with experienced drivers, novice drivers (less than 2000 km driving experience) were 30% less accurate and had significantly lower confidence scores for determining the meaning behind other drivers' formal signal use. Differences in comprehension between novice and experienced drivers were greatest for informal signals, such as flashing headlights. These findings suggest that comprehension of communication signals depend in part on driving experience.

The comprehension of interpersonal communications can also be influenced by the ways in which communication signals are used. For example, Takada, Fukuda, and Iwamiya (*14*) conducted a survey among 140 Japanese drivers and pedestrians on comprehension of horn use. The study found that horn signal comprehension varied depending on the temporal patterns of the signal (e.g., a single long honk vs. multiple short honks). Further, the study found even within a specific horn use pattern, there was general disagreement about the meaning of that signal. Additionally, pedestrians found car horns more startling and annoying than drivers and were less likely to understand the meaning behind the horn use, interpreting the horn use as primarily for gaining attention or to inform of danger even if the intended meaning was something different such as expressing gratitude or venting anger. The authors concluded that horns were not an effective means for interpersonal communication because they do not convey clear messages.

Summary and Discussion

This review synthesized the literature on roadway interpersonal communication with an objective of gaining a better understanding of the potential issues and challenges that may need to be addressed in the successful implementation of autonomous vehicles on public roadways. The review documents that interpersonal communication among road users not only takes places but is an important and understudied aspect of safe roadway travel.

A variety of nonverbal communication strategies are employed by drivers, pedestrians, and bicyclists ranging from the use of formal, technology-based strategies (such as turn signal use) to informal strategies (such as hand gestures or eye contact). While there is little research on the frequency with which these strategies are used, evidence suggests that they are commonly employed.

The review specifically addressed the interpersonal communication strategies utilized for communication between drivers and vulnerable road users (pedestrians and bicyclists), particularly for roadway crossings. The review documented the difficulties that vulnerable road users have in getting drivers to yield for roadway crossings and the important role interpersonal communication can play in increasing the chance of getting a driver to yield. Because vulnerable road users do not have technology-based strategies, they frequently employ body language, eye contact, and smiling to communicate to drivers their intent to cross a roadway. These strategies increase driver yielding behavior to some extent, but overall yielding is still low. Some research has investigated the use of hand gestures, which also moderately increase yielding behaviors.

Despite the variety of ways that road users engage in interpersonal communication, there are a surprising number of factors that can influence message comprehension, such as the context in which messaging takes place. An identical message, such as flashing head lights, often is comprehended differently for different road users because of these factors. One of these factors stems from the fact that many informal message strategies have developed out of necessity and are learned while people travel on the roadway. Thus, different areas of the United States and different countries have developed different interpersonal message norms that do not necessarily mean the same things. This also explains the fact that informal interpersonal roadway message comprehension is better among experienced drivers. Another factor affecting comprehension is that for some strategies the message's meaning changes depending on the context. For example, the flashing of headlights might have one meaning if you have just merged in front of a driver and another meaning if the driver has been following behind you.

These findings shed light on potential issues and challenges of interpersonal communication and the introduction of autonomous vehicles to the roadway, a topic that others have started to discuss (see e.g., *10*, *38*). These issues, challenges, and research needs are:

- Regardless of how quickly the vehicle fleet transforms to autonomous vehicles, the issues surrounding interpersonal communication will need to be addressed. Even with a fully autonomous fleet, there will still be vulnerable road users for which the autonomous vehicles will need to safely and effectively interact. Given the importance of interpersonal communication for roadway crossing events in particular, there will need to be intelligent strategies for the vulnerable roadway user to signal intent to cross and for the vehicle to signal whether or not it will yield. For example, some work is being done to address this. The Mercedes-Benz F 015 vehicle concept is designed with LED lights that communicate if the vehicle is in autonomous driving mode (*39*). The Nissan IDS concept has LED lights along its sides that shine a white light if pedestrians or bicyclists are nearby – to indicate that the vehicle is aware of their presence (*40*). The Rinspeed Oasis concept uses laser projection to the ground in front of the vehicle to warn pedestrians of its approach (*41*). In recognition of the research on facial expressions in pedestrian–vehicle interaction, Semcon's "smiling car" concept shows a curved line on an LED display when it yields to pedestrians (*42*).

- Autonomous vehicles will need to be able to not only detect but accurately interpret the variety of strategies used by all road users. This interpretation is nontrivial given the several factors that can affect comprehension, including cultural norms and the context in which the message-strategy is used. Thus, the sensing and control system for the autonomous vehicle will have to measure and analyze information in addition to the specific message-strategy.
- Autonomous vehicle algorithms that are designed to predict the intent of other drivers will have to rely on both formal and informal communication strategies. For example, the research on turn signal use shows that when used, turn signals do a good job of signaling the intent to turn, but they are not used all the time. Thus, a lack of turn signal use is not a sufficient predictor of whether or not a driver will turn.
- Traffic safety professionals might want to consider developing and promoting greater standardization among the various informal communication strategies, in much the same way as the arm signal to indicate the intent to turn left or right was developed. Gestures could be standardized for pedestrian crossing, increasing the autonomous vehicle's ability to correctly comprehend a pedestrian's intent.
- There may be value in developing educational courses for pedestrians and bicyclists on autonomous vehicles. Once autonomous vehicle design nears completion and becomes more standardized, courses can include topics on how autonomous vehicles work and how to interact with them.
- Interpersonal communication is a transaction among two or more parties. As such, autonomous vehicles will likely need a set of communication strategies that can be used to both acknowledge the reception of a message and to communicate the vehicle's intended behaviors to other road users.
- Along this same line, the research showed that drivers used relative speed as a means to communicate (e.g., slowing down or speeding up to indicate intent) with vulnerable road users. For autonomous vehicles to more effectively communicate with road users, they may need programming to incorporate this form of communication, both in terms of signaling intent and in receiving messages from other drivers. For example, Google programmed their autonomous vehicles to move slowly forward at 4-way stop sign intersections to indicate intent to proceed (*43*).
- An important medium for pedestrians to communicate to drivers is facial, socially based expressions such as smiling. In human-to-human contact, smiling occurs in the context of this social interaction. It is unlikely that this type of communication will naturally occur with autonomous vehicles present because pedestrians will not have another human to direct social-based expressions for communication to. Thus, one important medium of communication, particularly among vulnerable road users, may not be available for autonomous vehicles.
- Autonomous vehicle behavior will likely be predictable, and so road users may develop expectations of how autonomous vehicles will behave over time. Despite this, autonomous vehicles should still have a means to communicate intent, or at least inform road users of when they are in automated mode. Communication of intent may also help road users become more comfortable around autonomous vehicles. The effect of pedestrian experience with autonomous vehicles on crossing behavior should be studied further.
- The literature on interpersonal communication is sparse. To develop effective ways for autonomous vehicles to detect, receive, and properly interpret interpersonal communication on the roadway, there is a critical need to better understand how people do this effectively and when this communication fails. Further research is needed.

Acknowledgments

This research was supported by the Center for Advancing Transportation Leadership and Safety (ATLAS Center). The ATLAS Center is supported by a grant from the U.S. Department of Transportation, Office of the Assistant Secretary for Research and Transportation, University Transportation Centers Program (DTRT13-G-UTC54). The ATLAS Center is a collaboration between the University of Michigan Transportation Research Institute (UMTRI) and the Texas A&M Transportation Institute (TTI).

Author Contributions

The authors confirm contribution to the paper as follows: study conception and design: Stanciu, Eby; data collection: Stanciu; analysis and interpretation of results: Stanciu, Eby, Kostyniuk; draft manuscript preparation: Stanciu, Eby, Molnar, St. Louis, Zanier, and Kostyniuk. All authors reviewed the results and approved the final version of the manuscript.

References

1. Renge, K. Effect of Driving Experience on Drivers' Decoding Process of Roadway Interpersonal Communication. *Ergonomics*, Vol. 43, No. 1, 2000, pp. 27–39. https://doi.org/10.1080/001401300184648.
2. Shor, R. E. Shared Patterns of Nonverbal Normative Expectations in Automobile Driving. *The Journal of Social Psychology*, Vol. 62, No. 1, 1964, pp. 155–163. https://doi.org/10.1080/00224545.1964.9919512.
3. Zaidel, D. M. A Modeling Perspective on the Culture of Driving. *Accident Analysis & Prevention*, Vol. 24, No. 6, 1992, pp. 585–597. https://doi.org/10.1016/0001-4575(92)90011-7.
4. Ba, Y., W. Zhang, B. Reimer, Y. Yang, and G. Salvendy. The Effect of Communicational Signals on Drivers' Subjective Appraisal and Visual Attention during Interactive Driving Scenarios. *Behaviour & Information Technology*, Vol. 34, No. 11, 2015, pp. 1107–1118. https://doi.org/10.1080/0144929X.2015.1056547.
5. Sullivan, J. M., S. Bao, R. Goudy, and H. Konet. Characteristics of Turn Signal Use at Intersections in Baseline Naturalistic Driving. *Accident Analysis & Prevention*, Vol. 74, 2015, pp. 1–7. https://doi.org/10.1016/j.aap.2014.10.005.
6. Drivemocion. http://www.drivemocion.com/. Accessed January 25, 2017.
7. National Highway Traffic Safety Administration. *National Motor Vehicle Crash Causation Survey: Report to Congress*. Publication DOT HS 811 059. DOT HS, U.S. Department of Transportation, 2008, p. 059.
8. Iliaifar, A. NHTSA: Autonomous Car Technology Could Eliminate up to 80 Percent of Crashes. *Digital Trends*, April 27, 2012. http://www.digitaltrends.com/cars/nhtsa-autonomous-driving-cars-could-eliminate-up-to-80-percent-of-crashes/. Accessed Jan. 25, 2017.
9. Ziegler, C. A Google Self-Driving Car Caused a Crash for the First Time. *The Verge*, 2016. http://www.theverge.com/2016/2/29/11134344/google-self-driving-car-crash-report. Accessed January 16, 2017.
10. Parkin, J., B. Clark, W. Clayton, M. Ricci, and G. Parkhurst. *Understanding Interactions between Autonomous Vehicles and Other Road Users: A Literature Review*. University of the West of England, Bristol, 2016.
11. National Highway Traffic Safety Administration. *Fatality Analysis Reporting System (FARS)*. https://www.nhtsa.gov/research-data/fatality-analysis-reporting-system-fars. Accessed January 25, 2017.
12. Renge, K., G. Weller, B. Schlag, M. Peräaho, and E. Keskinen. Comprehension and Evaluation of Road Users' Signaling - An International Comparison between Finland, Germany and Japan. In *Traffic and Transport Psychology: Theory and Application: Proceedings of the ICTTP 2000* (T. Rothengatter and R. D. Huguenin, eds.), Elsevier, Oxford, pp. 91–100.
13. Mesken, J. *Measuring Emotions in Traffic*. Publication D-2002-3. SWOV Institute for Road Safety Research, Leidschendam, The Netherlands, 2002.
14. Takada, M., Y. Fukuda, and S.-I. Iwamiya. Questionnaire Survey on Vehicle Horn Use. *Journal of the Acoustical Society of America*, Vol. 123, No. 5, 2008, p. 3260.
15. Faw, H. W. To Signal or Not to Signal: That Should Not Be the Question. *Accident Analysis & Prevention*, Vol. 59, 2013, pp. 374–381. https://doi.org/10.1016/j.aap.2013.06.036.
16. Šucha, M. Road Users' Strategies and Communication: Driver-Pedestrian Interaction. Presented at 5th Conference Transport Solutions from Research to Deployment, Transport Research Arena, Paris, 2014.
17. Crowley-Koch, B. J., R. Van Houten, and E. Lim. Effects of Pedestrian Prompts on Motorist Yielding at Crosswalks. *Journal of Applied Behavior Analysis*, Vol. 44, No. 1, 2011, pp. 121–126. https://doi.org/10.1901/jaba.2011.44-121.
18. Zhuang, X., and C. Wu. Pedestrian Gestures Increase Driver Yielding at Uncontrolled Mid-Block Road Crossings. *Accident Analysis & Prevention*, Vol. 70, 2014, pp. 235–244. https://doi.org/10.1016/j.aap.2013.12.015.
19. Turnbaugh, W. A., and S. P. Turnbaugh. American Greetings: Hand Signaling on the Highways. *Symbolic Interaction*, Vol. 10, No. 1, 1987, pp. 139–142. https://doi.org/10.1525/si.1987.10.1.139.
20. Böckler, A., R. P. R. D. van der Wel, and T. N. Welsh. Catching Eyes Effects of Social and Nonsocial Cues on Attention Capture. *Psychological Science*, 2014, pp. 720 –727. https://doi.org/10.1177/0956797613516147.
21. Guéguen, N., S. Meineri, and C. Eyssartier. A Pedestrian's Stare and Drivers' Stopping Behavior: A Field Experiment at the Pedestrian Crossing. *Safety Science*, Vol. 75, 2015, pp. 87–89. https://doi.org/10.1016/j.ssci.2015.01.018.
22. Bertulis, T., and D. Dulaski. Driver Approach Speed and Its Impact on Driver Yielding to Pedestrian Behavior at Unsignalized Crosswalks. *Transportation Research Record: Journal of the Transportation Research Board*, 2014. 2464: 46–51.
23. Emerson, R. W., E. Bourquin, D. Sauerburger, and J. Barlow. *Conditions That Influence Drivers' Yielding Behavior at Uncontrolled Crossings and Intersections with Traffic Signal Controls*. Publication TRCLC 14-04. Western Michigan University, Kalamazoo, MI, 2015.
24. Foster, N., C. Monsere, and K. Carlos. Evaluating Driver and Pedestrian Behaviors at Enhanced, Multilane, Midblock Pedestrian Crossings. *Transportation Research Record: Journal of the Transportation Research Board*, 2014. 2464: 59–66.
25. Ibrahim, N. I., F. A. Kidwai, and M. R. Karim. Motorists and Pedestrian Interaction at Unsignalised Pedestrian Crossing. *Proceedings of the Eastern Asia Society for Transportation Studies*, Vol. 5, 2005, pp. 120–125.
26. Stapleton, S., T. Kirsch, T. J. Gates, and P. T. Savolainen. Factors Affecting Driver Yielding Compliance at Uncontrolled Midblock Crosswalks on Low-Speed Roadways. *Transportation Research Record: Journal of the Transportation Research Board*, 2017. 2661: 95–102.
27. Lehsing, C., T. Benz, and K. Bengler. Insights into Interaction - Effects of Human-Human Interaction in Pedestrian Crossing Situations Using a Linked Simulator Environment. *IFAC-PapersOnLine*, Vol. 49, No. 19, 2016, pp. 138–143. https://doi.org/10.1016/j.ifacol.2016.10.475.
28. Nasar, J. L. Prompting Drivers to Stop for Crossing Pedestrians. *Transportation Research Part F: Traffic Psychology and Behaviour*, Vol. 6, No. 3, 2003, pp. 175–182. https://doi.org/10.1016/S1369-8478(03)00024-X.
29. Guéguen, N., C. Eyssartier, and S. Meineri. A Pedestrian's Smile and Drivers' Behavior: When a Smile Increases Careful Driving. *Journal of Safety Research*, Vol. 56, 2016, pp. 83–88. https://doi.org/10.1016/j.jsr.2015.12.005.

30. Schneemann, F., and I. Gohl. Analyzing Driver-Pedestrian Interaction at Crosswalks: A Contribution to Autonomous Driving in Urban Environments. *Proc., 2016 IEEE Intelligent Vehicles Symposium (IV)*, Gothenburg, Sweden, 2016.
31. Clamann, M., M. Aubert, and M. L. Cummings. Evaluation of Vehicle-to-Pedestrian Communication Displays for Autonomous Vehicles. Presented at 96th Annual Meeting of the Transportation Research Board, Washington, D.C., 2017.
32. Walker, I. Signals are Informative but Slow Down Responses When Drivers Meet Bicyclists at Road Junctions. *Accident Analysis & Prevention*, Vol. 37, No. 6, 2005, pp. 1074–1085. https://doi.org/10.1016/j.aap.2005.06.005.
33. Walker, I., and M. Brosnan. Drivers' Gaze Fixations during Judgements about a Bicyclist's Intentions. *Transportation Research Part F: Traffic Psychology and Behaviour*, Vol. 10, No. 2, 2007, pp. 90–98. https://doi.org/10.1016/j.trf.2006.06.001.
34. Archer, D. Unspoken Diversity: Cultural Differences in Gestures. *Qualitative Sociology*, Vol. 20, No. 1, 1997, pp. 79–105. https://doi.org/10.1023/A:1024716331692.
35. Factor, R., D. Mahalel, and G. Yair. The Social Accident: A Theoretical Model and a Research Agenda for Studying the Influence of Social and Cultural Characteristics on Motor Vehicle Accidents. *Accident Analysis & Prevention*, Vol. 39, No. 5, 2007, pp. 914–921. https://doi.org/10.1016/j.aap.2006.12.015.
36. McGarva, A. R., and M. Steiner. Provoked Driver Aggression and Status: A Field Study. *Transportation Research Part F: Traffic Psychology and Behaviour*, Vol. 3, No. 3, 2000, pp. 167–179. https://doi.org/10.1016/S1369-8478(00)00023-1.
37. Novaco, R. W. Aggression on Roadways. In *Advances in Psychology* (R. Baenninger, ed.), North-Holland, pp. 253–326.
38. Köhler, S., M. Goldhammer, S. Bauer, K. Doll, U. Brunsmann, and K. Dietmayer. Early Detection of the Pedestrian's Intention to Cross the Street. *Proc., 2012 15th International IEEE Conference on Intelligent Transportation Systems*, New York, NY, 2012.
39. Mercedes-Benz. The Mercedes-Benz F 015 Luxury in Motion. *Mercedes-Benz Press Releases*, January 2015. https://www.mbusa.com/mercedes/about_us/press/pressId-ad64f6d63f-6da410VgnVCM100000ccec1e35____. Accessed November 1, 2017.
40. Nissan. Nissan IDS Concept: Nissan's Vision for the Future of EVs and Autonomous Driving. *Nissan Online Newsroom*, October 2015. http://nissannews.com/en-US/nissan/usa/releases/nissan-ids-concept-nissan-s-vision-for-the-future-of-evs-and-autonomous-driving. Accessed November 1, 2017.
41. Hull, R. Is it a Car or a Greenhouse? Rinspeed's New Vehicle Has a Steering Wheel that Doubles as a Table and a Garden for Growing Radishes behind the Windscreen. *This is Money*, December 2016. http://www.thisismoney.co.uk/~/article-4009296/index.html. Accessed November 1, 2017.
42. Semcon. The Smiling Car - Who Sees You When the Car Drives Itself? *Semcon*. https://semcon.com/smilingcar/. Accessed November 1, 2017.
43. Naughton, K. Humans are Slamming Into Driverless Cars and Exposing a Key Flaw. *Bloomberg Technology*, December 2015. https://www.bloomberg.com/news/articles/2015-12-18/humans-are-slamming-into-driverless-cars-and-exposing-a-key-flaw. Accessed October 26, 2017.

The Standing Committee on Vehicle-Highway Automation (AHB30) peer-reviewed this paper (18-03038).

The contents of this paper reflect the views of the authors, who are responsible for the facts and the accuracy of the information presented. This paper is disseminated under the sponsorship of the U.S. Department of Transportation's University Transportation Centers Program, in the interest of information exchange. The U.S. Government assumes no liability for the contents or use thereof.

Article

TRR
JOURNAL OF THE TRANSPORTATION RESEARCH BOARD

Transportation Research Record
2018, Vol. 2672(22) 67–77

DOI: 10.1177/0361198118796375
journals.sagepub.com/home/trr
SAGE

On Augmenting Adaptive Cruise Control Systems with Vehicular Communication for Smoother Automated Following

Jan-Niklas Meier[1], Aravind Kailas[2], Oubada Abuchaar[3], Maher Abubakr[4], Rawa Adla[5], Mahdi Ali[6], George Bitar[7], Richard Deering[8], Umair Ibrahim[9], Paritosh Kelkar[10], Vivek Vijaya Kumar[11], Ehsan Moradi-Pari[10], Jay Parikh[8], Samer Rajab[10], Masafumi Sakakida[12], and Masashi Yamamoto[12]

Abstract

This paper focuses on evaluating, in a structured manner, the potential benefits, along with the implementation and performance issues, of utilizing dedicated short range communication-based communication in conjunction with adaptive cruise control (ACC) systems. This work was done in the United States under a cooperative agreement between the Crash Avoidance Metrics Partners LLC and the Federal Highway Administration. Designing cooperative adaptive cruise control (CACC) as an extension of ACC, and by using a combination of a comprehensive simulation framework and test vehicles, benefits of vehicular communication on string stability were established, and the performance of the novel CACC-enabling software modules were validated. Another key contribution of this work is the consideration of vehicles with different dynamic responses as a part of a single string. Four light-duty vehicles (hatchback, mid- and full-size sedans, large SUV), each from a different automotive original equipment manufacturer, were retrofitted with common ACC and vehicular communication systems. They were tested under many different conditions to obtain performance data (such as radar sensor readings, etc.) when operating in a vehicle string. These data were then integrated into the simulation environment to develop and validate the CACC modules. The paper concludes with a recommendation of some data elements for over-the-air messages to enable CACC functionality.

Cooperative adaptive cruise control (CACC) is an application that uses longitudinal control to improve traffic flow and driving convenience, and, when used in conjunction with supporting technologies, offers the potential to reduce the congestion on most urban freeways and arterials. The CACC system adapts inter-vehicle distances by leveraging real-time information from vehicles in close proximity along with local sensor measurements. In theory, improving the traffic flow stability can lead to improved fuel economy (or energy savings) (*1*). The United States Department of Transportation has actively engaged in assessing applications that realize the full potential of connected vehicles, travelers, and infrastructures to enhance current operational practices and transform future surface transportation systems management through connected vehicles. Prior Federal Highway Administration-sponsored research suggests a key benefit of implementing CACC could be attaining traffic flow stability as a result of dynamic and automatic modulation of cruise control speeds of a "string" of vehicles (*1*). Arnaout and Bowling also highlight the improved throughput on a given road segment as a result of more efficient vehicle distribution and consistent flow of traffic when using CACC (*2*). Other benefits of CACC include higher fuel efficiency resulting from the increased aerodynamic efficiency provided by decreasing gaps between

[1]Volkswagen Group of America, Auburn Hills, MI
[2]Volvo Group North America, Costa Mesa, CA
[3]Representing: Self, Farmington Hills, MI
[4]SUBARU Research & Development, Inc., Ann Arbor, MI
[5]Ford Motor Company, Farmington Hills, MI
[6]Hyundai-Kia America Technical Center, Inc., Superior Township, MI
[7]Volvo Group North America, Greensboro, NC
[8]Representing: CAMP LLC, Farmington Hills, MI
[9]Ford Motor Company RIC-PAW, Palo Alto, CA
[10]Honda R&D Americas, Inc., Southfield, MI
[11]General Motors, Warren, MI
[12]Mazda North American Operations, Irvine, CA

Corresponding Author:
Address correspondence to Aravind Kailas: aravind.kailas@volvo.com

string members (*3*) and from reducing instances of unnecessary braking / throttle maneuvers (*4*). From reviewing past and on-going CACC works in Parikh et al. (*5*), a broad, high-level research plan was developed to identify the potential benefits, opportunities, safety issues, technical gaps, and challenges in deploying CACC systems. In addition, a more focused set of recommendations was developed for production implementation of CACC in future vehicles that included a prototype and small-scale test plans. This work keys off an analysis of practical highway behaviors for the real-world situation in which different brands of vehicles with different characteristics comprise a vehicle string and interact with each other, and dedicated short range communication (DSRC)-based infrastructures. Under such conditions, the formed group exhibits car-following and lane-changing behaviors. Although improving string stability was the focus of the majority of prior works considering homogenous (*6–10*) and heterogeneous (*11*) string vehicles, the unavailability of a reference production adaptive cruise control (ACC) system limited the research to using identical vehicles for experiments. In addition, with the exception of Lu et al. that concerns three trucks of the same configuration (*12*), none of other prior works consider looking ahead more than one vehicle in the formed string using DSRC. Furthermore, they do not consider common driving maneuvers such as highway merge, braking, and stationary obstacle avoidance. The authors identified no single experiment which implemented heterogeneous string formation by designing a reference standard ACC system and created a virtual target to take advantage of range information obtained from DSRC, including multi-vehicle look-ahead, in order to observe system behavior in multiple situations.

With this in mind, the work focused on the enhancement of radar-based, production ACC systems using DSRC-based vehicle-to-vehicle (V2V) / vehicle-to-infrastructure (V2I) communication and messages. Baseline performance was established using road test data where a "reference" ACC and DSRC-based V2V communication were systems operating concurrently. This included, for example, an evaluation of response times during longitudinal maneuvers (i.e., braking and throttle) in vehicle-following scenarios. The road test data were utilized to parameterize the software-based simulator. The simulator in turn was used to design, test, and evaluate the CACC algorithms and their applicability. Further analysis of the gathered data and research findings will provide direct support for the development of technical standards such as the SAE J2945/6 that enables CACC and platooning services.

This paper is organized as follows. Project Overview describes the research project background, followed by a description of the CACC-enabling modules. The key findings from the road tests and simulation-based experiments are presented in the Research Findings section. Finally, concluding remarks and suggestions for future work complete the remainder of the paper.

Project Overview

The overarching goal of the project was to investigate the feasibility of a CACC prototype in a systematic manner. This was accomplished by testing a common ACC implementation on the roads and incorporating empirical data to parameterize a simulation environment for developing and analyzing the performance of novel CACC modules developed in the project. Testing on the roads was conducted to collect both radar sensor and V2V data to gain insight into the baseline performance of multi-vehicle strings of autonomous ACC systems under certain conditions. Evaluation of these data helped to identify potential opportunities to expand functionality using CACC. The simulation environment comprised of communications and sensor models, and a microscopic traffic simulator to simulate vehicle interactions, as well as the impact of these vehicles on the overall traffic flow.

Four vehicles (hatchback, mid-size and full-size sedans, and a large SUV) from different manufacturers were used in the road testing to capture the impacts of different original equipment manufacturers (OEMs) vehicle control interfaces in addition to the differences in longitudinal control system design. All vehicles contained production ACC systems capable of stop-and-go operation. So, the required control systems and actuators needed to implement the experimental ACC / CACC system were already installed. A common ACC system (called "reference" ACC system) was developed and implemented on all four vehicles. This reference ACC system, and not the production ACC systems, was used during the road tests. The design choices were motivated by the prospects of defining a common design platform and subsequently augmenting the developed CACC functionalities in a structured manner. The diverse design choices of the vehicle control systems by the respective OEMs were identified during vehicle integration. The rationale for a common ACC system on all test vehicles was to tap into the native responses of the hardware system in each vehicle. But despite this choice, significant variations in the system performance / responses were observed during the road-based tests.

The CACC capabilities were developed using the simulation environment that included a traffic simulator, radar, Global Positioning System (GPS), wireless communications models, and a vehicle dynamics model considering the different vehicle types. Multiple vehicles types were also simulated to assess the impacts of differences in vehicle sizes and dynamics on the system performance. Three broad capabilities for CACC were investigated. They are as follows:

1. Situational awareness enhancement: Increased awareness range (communication range compared with onboard sensors sensing range), wider field-of-view (immediately preceding the vehicle known as the "target vehicle" (TV), multi-vehicle ahead in the same and different lanes), and feasibility of providing additional information through communication to a longer and wider range are three immediate advantages of the CACC over the ACC. This research investigated whether the reduction in unnecessary acceleration / deceleration (i.e., smoother controller performance) and perturbations in surrounding traffic may be accomplished by improving the world view of the system. Specific examples are listed below:
   a) If the CACC string is approaching a group of slower vehicles, deceleration may begin earlier and more smoothly.
   b) If a vehicle ahead in the string is braking, deceleration may begin in anticipation prior to / in coordination with the response of the TV.
   c) If a vehicle in an adjacent lane activates its turn signal, the impending lane change can be anticipated and string behavior adapted more rapidly.

   Therefore, improving longitudinal control performance within the string by providing ACC with additional context information through communication was identified and considered as the first major capability of CACC.
2. System latency reduction: "Phantom" traffic jams are a result of perturbations in unstable vehicle strings. Typically, a string of ACC vehicles is not string-stable. According to Michaelian and Browand, the CACC system was strictly L2 string-stable, whereas the ACC system was not (*3*). This is because of the high rate of overshoot for the following vehicles. When a reduction in speed occurs within the string, the deceleration required by following vehicles increases for each following vehicle, eventually bringing the entire string to a full stop. Communicating the behavior of preceding vehicles, actual or planned, within the string might reduce response time and improve the string stability. Therefore, reducing the latency of the string response to perturbations to provide smoother, more consistent traffic flow leading to increased throughput was investigated.
3. Time-gap optimization: Time-gap is defined as the time needed by the host vehicle (HV) to travel the vehicle clearance (the distance between HV and the TV). The time-gaps for enhancing traffic throughput may not always be the obvious choice because of the traffic randomness and safety implications of following too closely. This motivated an investigation of the appropriate time-gaps for vehicles in a string. Specifically, reducing the time-gap within a string (to allow for an increase in vehicle throughput) while at the same time maintaining string stability (to dampen traffic perturbations) was set as a goal for the algorithm design.

CACC-Enabling Modules

Several unique software modules were developed to enhance the situational awareness and response times of the baseline system. This was motivated by the potential extensions to existing ACC longitudinal controllers and the prospect of an SAE J2945/1Basic Safety Message (BSM) (*13*) deployment. This was enabled by additional data elements that were received from preceding vehicles. Whereas relative distance and speed were key controller inputs during ACC operation, additional data such as the measured acceleration and "acceleration forecasts" are used during the CACC mode. Three major CACC-enabling concepts are described below.

Extending ACC through the Virtual Target Creation

A fully redesigned controller that leveraged the additional available information in CACC provides the most benefits (*14*), and could be considered for future production implementations. However, by adjusting the input data to an unmodified controller, the benefits of CACC were realized in terms of a dynamic system response.

During ACC operation, the controller receives the relative distance and speed of the TV while in CACC mode, and additional data, such as current measured acceleration and acceleration forecasts, are incorporated into the calculation. To this end, the concept of a "virtual target" was introduced. The Virtual Target Creation module receives the aforementioned information (from the TV) to predict the TV motion and inputs it to the longitudinal controller instead of the current speed and distance information. Based on current vehicle dynamics of both the HV and TV, the resulting "virtual target" can be closer or further away than the actual physical location of the TV. As a result, earlier reactions can be triggered in the longitudinal controller, thereby reducing the overall reaction time. In addition, this approach minimizes the required modifications on the existing ACC longitudinal controller that are already parameterized for driver comfort with a consideration of the

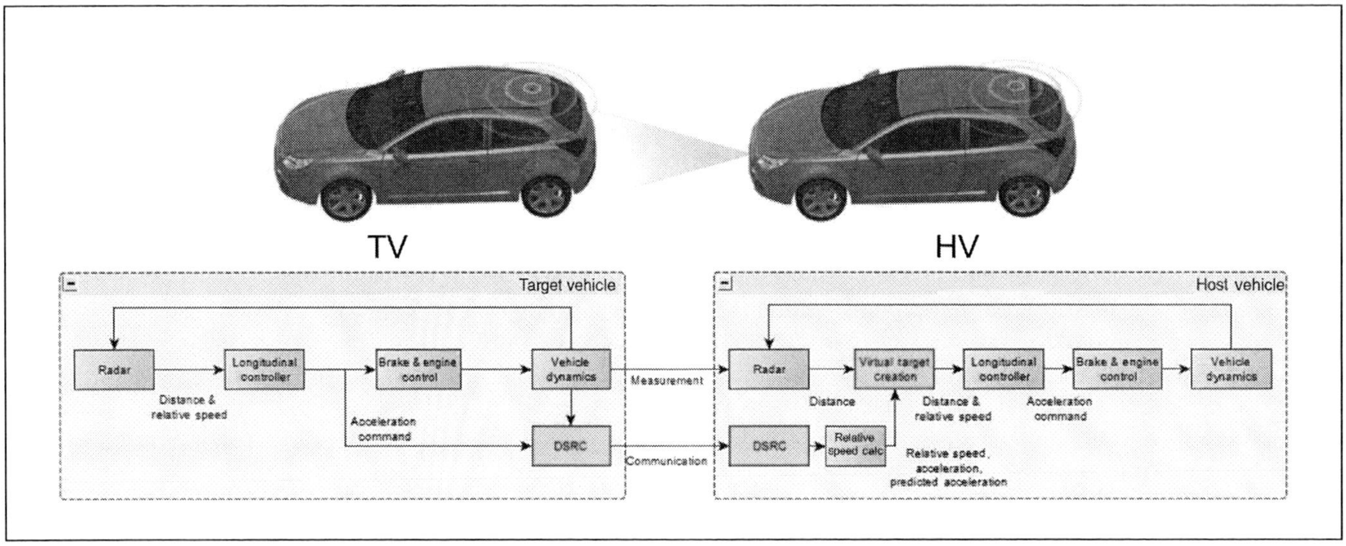

Figure 1. Control diagram illustrating the integration of the virtual target data creation module with the longitudinal controller between the target vehicle (TV) and host vehicle (HV).

inherent vehicle characteristics. Figure 1 illustrates the control diagram of two vehicles in the CACC string. The blue components represent an ACC control diagram, and the green components were added to enable CACC functionalities.

Vehicle Look-Ahead Concepts

The CACC-based longitudinal control implementation can operate in many modes. They are described below.

The one-vehicle look-ahead (OVLA) (illustrated in Figure 2) is the most basic implementation of CACC control. It is similar to the conventional ACC system where only data from the TV will be used in the decision-making, and the difference being that both radar and V2V data are used in the case of CACC. The obvious limitation is the propagation of information about perturbations along the stream in a sequential manner. Nevertheless, the string stability will be better than an ACC system.

The multi-vehicle look-ahead (MVLA) mode utilizes the V2V data from several (if not all) preceding vehicles in the string. It computes the "virtual" target by considering V2V data from multiple preceding vehicles and weighting them based on distance, thereby enhancing the HV reaction times to traffic perturbations.

The vehicle behavior estimation feature can operate in parallel with either of the previously mentioned modes. It assesses information received from preceding vehicles for potentially hazardous conditions such as harsh braking maneuvers. If such conditions are detected, it increases the target time-gap and triggers an initial deceleration.

In-Lane Assessment of Remote Vehicles

This CACC-based longitudinal control implementation incorporates the data received from preceding vehicles in the same lane. This necessitates the HV to identify and select the remote vehicles in its lane. In contrast, the conventional ACC systems conduct this assessment using radars that have a relatively high lateral accuracy. Physical locations of the remote vehicles can be obtained from basic safety messages (BSMs) transmitted and received by the vehicular communication systems. However, because these physical coordinates are based on relative GPS information, the desired accuracy and granularity (e.g., for a lane-level assessment) in certain situations may be insufficient. Therefore, the CACC implementation will consider both the radar sensor and V2V data.

From each DSRC-equipped vehicle in the line-of-sight of the radar, the HV receives two independent sets of readings (from radar and V2V unit). Typically, these readings are not identical and the object fusion module monitors both sets of readings over time and compares the relevant parameters (e.g., size, speed, distance, etc., to name a few). The sensor fusion module enables the HV to correlate the independent data to ascertain that they are from the same remote vehicle, thereby assessing whether the remote vehicle is in its lane. The "temporary ID" data element (present in the received BSM) of the remote vehicle is then recorded by the HV. This in-lane assessment approach is sufficient even when the lane association for the remote vehicle is unclear from the V2V data. Any subsequent transmissions from this remote vehicle will be incorporated in the decision-making as a result of the established "trust." The process

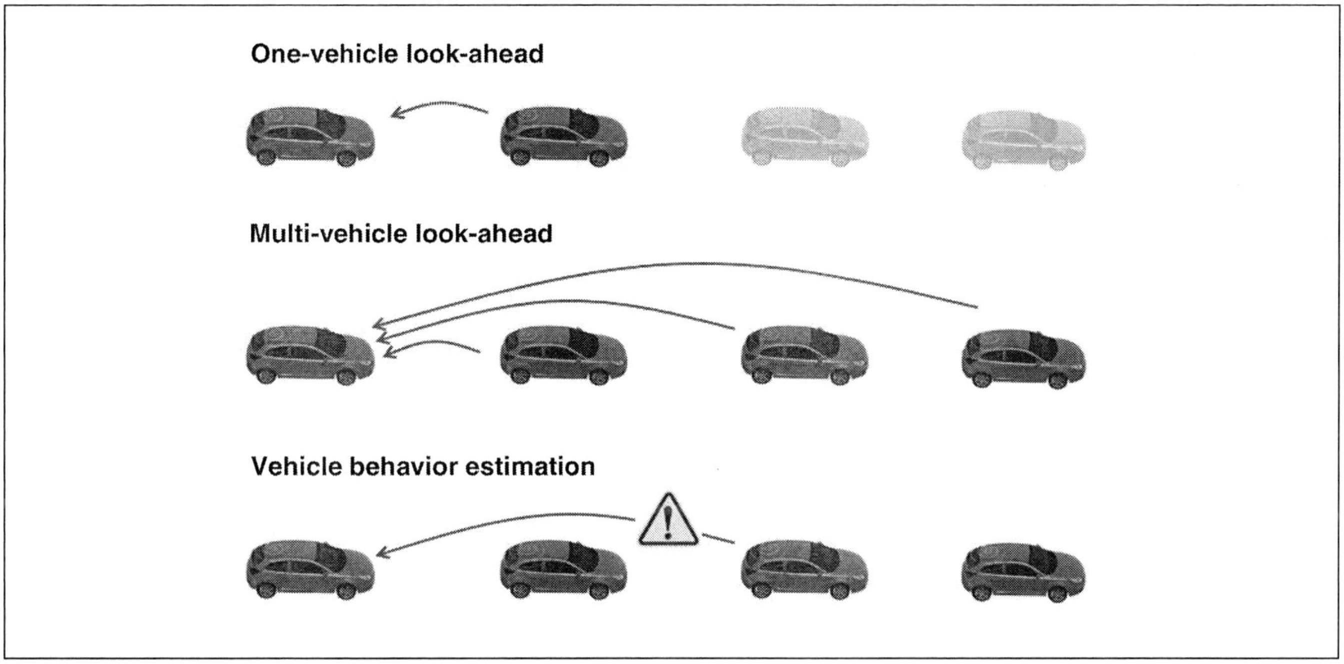

Figure 2. Illustration of the different implemented look-ahead features.

Figure 3. An illustration of in-lane assessment of a remote vehicle (the white vehicle) by a HV (the blue vehicle) by leveraging information from a preceding in-lane vehicle (the red vehicle).

of fusing the radar and V2V data is relatively simple and enables the OVLA CACC.

In the case of MVLA CACC, information, from not only the TV but also from the other preceding neighboring vehicles, will be utilized for in-lane assessment. Given that the radar sensor cannot reliably detect multiple vehicles ahead in the same lane, V2V data alone are utilized for in-vehicle assessment of remote vehicles. As alluded to earlier, the GPS accuracy may be compromised in certain situations, thereby rendering the GPS coordinates in the BSM (i.e., V2V data) unreliable. For instance, as illustrated in Figure 3, the radar view of the blue vehicle (considered to be the HV in this example) is obstructed by the red vehicle, and it may be unable to ascertain the lane association for the white vehicle based on V2V data alone, especially if the GPS coverage is choppy. This paper's novel approach for addressing this issue is to tap into the verification capability of an intermediate vehicle in its lane. By leveraging the "trust" established between the vehicles in the same lane, once the red vehicle has validated the white vehicle to be in its lane, this information can be passed along to the blue vehicle, thereby enabling the HV to reliably carry out in-lane assessment of the white vehicle.

Therefore, a requirement was added that each vehicle will transmit its own temporary ID and that of its TV. By

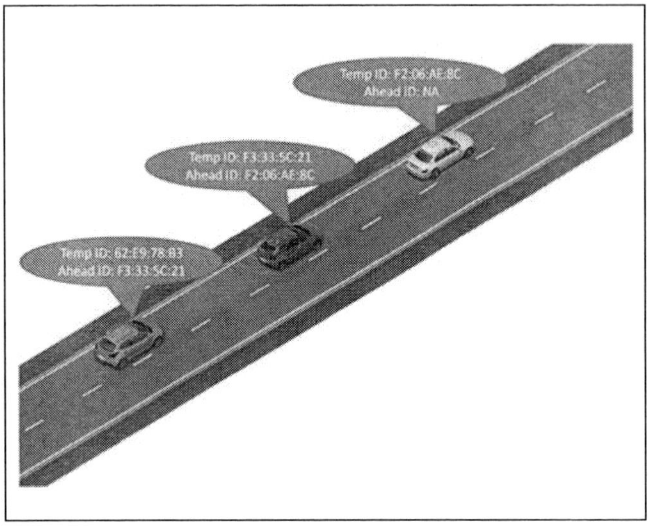

Figure 4. The "Ahead ID" for the white car is not available implying that it is the lead vehicle in the string, whereas the "Ahead ID" maintained by red car correctly corresponds to the temporary ID of the white car, the TV for the red car.

utilizing this information, vehicles can build on-the-fly a linked list of vehicles in the same string through received BSMs from vehicles ahead and identifying one link after another until either communication range is exceeded or the beginning of the string is reached, as illustrated in Figure 4.

In the event of vehicle cut-ins and cut-outs, the involved vehicles will update their current TV information and this information will propagate to other vehicles keeping track of string membership.

In addition, as the algorithm requires message reception from multiple vehicles ahead, it must anticipate that the HV will not receive every single V2V message that is being transmitted from vehicles further ahead. Therefore, the algorithm needs to be robust to packet drops, and this can be achieved by implementing an "information age threshold" (not described in this paper) to determine its expiration.

Research Findings

The research utilized a combination of both road-based and simulator-based tests to establish the performance of the newly designed CACC modules. The common ACC system implementation on all four test vehicles was evaluated using road tests under a variety of scenarios. A total of 25 different scenarios were considered to characterize ACC system behavior and parameterize vehicle dynamics for the simulation environment. Simulations were used to evaluate and develop CACC algorithms and to compare them with the ACC baseline performance. The environment included a microscopic traffic simulator (VISSIM), wireless communication models (NS3) as well as models for vehicle dynamics, GPS, and radar (implemented in MATLAB / Simulink). The simulation models were parameterized based on the road test data that were collected using the four project vehicles. The simulation environment was then used to implement, debug, and evaluate the CACC modules, and establish its efficacy relative to ACC in three broad areas: situational awareness, system latency, and time-gap. The findings are discussed below.

Enhancements to Situational Awareness of a CACC System

The simulations indicated that the MVLA CACC approach, in combination with behavior estimation, enhanced the situational awareness tremendously as compared with CACC with OVLA. The simulation scenario considered a five-vehicle CACC string comprising three compact cars followed by two large SUVs that have relatively slow dynamics. The following time-gap was chosen to be 0.5 s. A perturbation (acceleration / deceleration) modeled as a sine wave was introduced in the string. Figure 5 is a truncated plot of temporal variations in the accelerations of the vehicles in the CACC string under the OVLA and MVLA approaches. From the plot, it can be observed that the peak-to-peak value of the accelerations for the last two vehicles (the SUVs) in the string is smaller by 16% in the case of MVLA. In other words, the benefits of MVLA seem to be more pronounced for the vehicles with slower dynamic responses, that is, the fourth and fifth vehicles in the current set up.

Reduction in System Latency

To study the reaction times of the simulated system, separate strings of ACC and CACC vehicles were subjected to a sequence of positive and negative accelerations of $2.5\,m/s^2$, and the reaction times to braking in preceding vehicles were recorded. The input perturbation is introduced in the first vehicle of the string. To study system latency, a five-vehicle string with all hatchback vehicles was considered. For this scenario, five same-model vehicles were used because that removed the influence of five different models of vehicle, each with a different native response. The first vehicle in Figure 5 (blue lines) was artificially controlled and the time-gap for following vehicles was set to 1 s for both ACC and CACC to allow for a direct comparison. For CACC, only the OVLA mode was used for a better direct comparison of the reaction times. The MVLA mode would have reduced latencies further as a result of the decrease or removal of propagation time through the string.

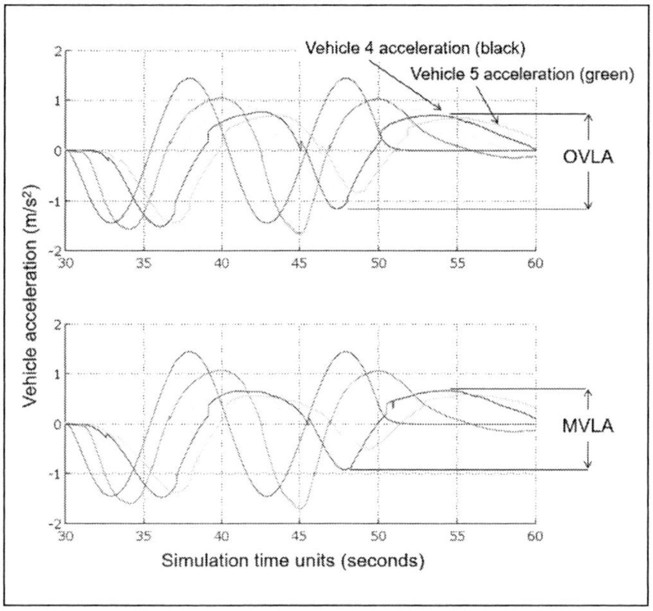

Figure 5. Temporal variations in the accelerations of the vehicles in the CACC string are shown for one-vehicle look-ahead (OVLA) and multi-vehicle look-ahead (MVLA). The accelerations of the five vehicles are denoted by five different colors, with the vehicles 4 and 5 represented by black and green curves, respectively.

Figure 6, *a* and *b* show the simulated acceleration responses of ACC and CACC strings, respectively. In both plots, the dashed lines represent the acceleration commands of a vehicle and the solid lines represent the actual physical acceleration of the vehicle. Let us consider the ACC string first. As can be seen from Figure 6*a*, the ACC vehicles sequentially decelerate quickly with each vehicle in the string reaching the maximum pre-set deceleration limit of $3\,m/s^2$. After the vehicles decelerate, they accelerate again to close the increased time-gap, implying that the applied deceleration was perhaps too harsh. On the other hand, as can be seen in Figure 6*b*, the CACC vehicles start decelerating significantly earlier, and, therefore, do not reach the pre-set limit of the controller. Furthermore, with the exception of the second vehicle, there is no acceleration overshoot after the deceleration maneuver has been completed. The CACC string becomes stable around the 34th time instance, whereas the downstream vehicles in the ACC string are significantly unstable even beyond the 40th time instance (as indicated in the dash–dot box). The slight overshoot observed in the CACC response can be attributed to the suddenness in the perturbation and the inherent operation of the longitudinal controller that limits how fast a certain deceleration value can be reached.

The propagation delays associated with the deceleration / acceleration maneuvers of the vehicles in the string can be divided into:

(a) A "detection delay," which is defined as the time between the HV detecting an acceleration / deceleration in a preceding vehicle and issuing the throttle / brake commands.
(b) A "reaction delay," which is defined as the time between the HV experiencing the throttle / brake response after issuing the commands.

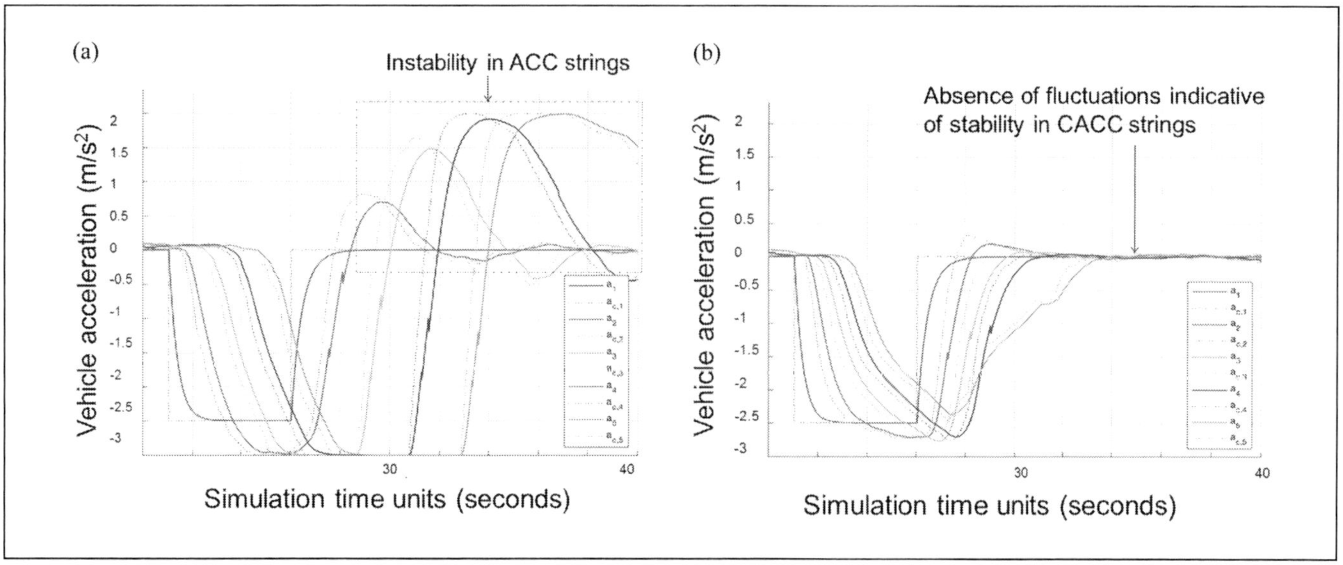

Figure 6. The time required for string stability in the event of a perturbation is much less in the case of cooperative adaptive cruise control (CACC) as compared with adaptive cruise control (ACC). This is denoted by the absence of fluctuations around the 34th second for the CACC string in (*b*).

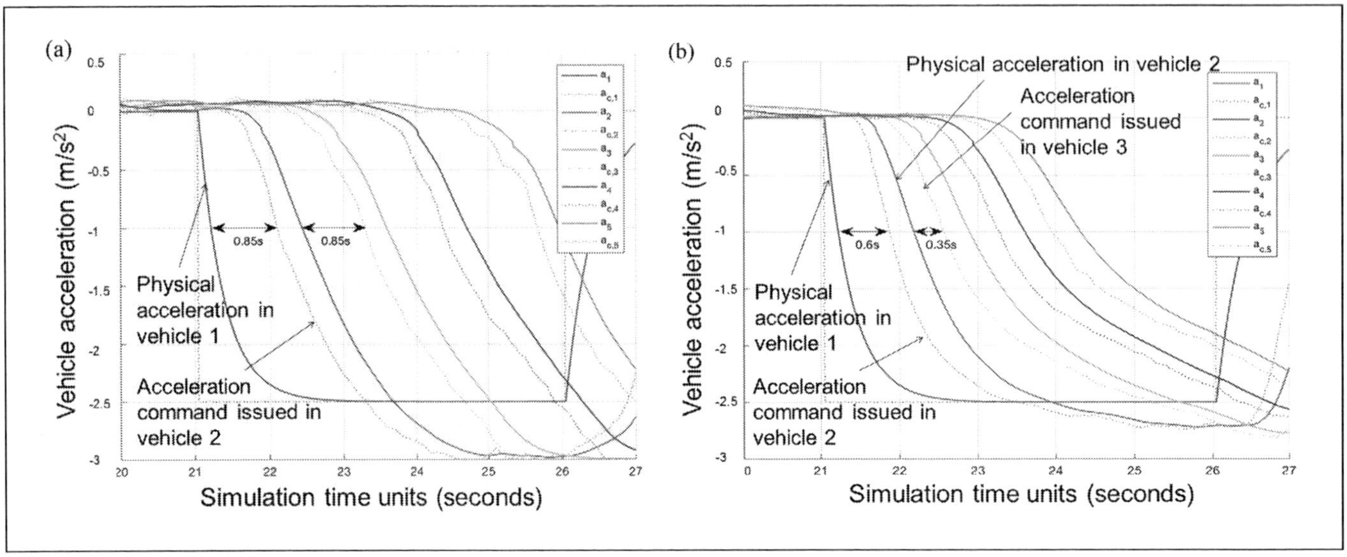

Figure 7. The detection delays for (a) ACC in and (b) OVLA CACC in. From (*b*), two things are obvious: (1) a reduction in detection delay in the case of CACC; (2) faster detection of the braking maneuver in a preceding vehicle as a result of acceleration commands transmissions.

The reaction delay is a function of the vehicle powertrain, and optimizing it is outside the scope of the research work. As a result, there are no noticeable differences between simulation-based results for ACC and CACC. So, in other words, only the *detection delays* are considered in the analysis.

In Figure 7, the dashed lines represent the acceleration commands of a vehicle and the solid lines represent the actual physical acceleration of the vehicle. Let us consider the ACC string. The detection delay for ACC was nearly identical for all vehicles in the string and about 0.85 s. The delay for the CACC vehicles differs between the first and the second vehicles (recorded as 0.6 s) and all the other vehicles in the string (0.35 s). This can be attributed to the first vehicle behaving like a manually operated, V2V-capable vehicle, in which it only transmits current acceleration values. Without any knowledge of future acceleration states (or "acceleration forecasts"), the second vehicle is unable to predict the behavior of its TV, that is, the first vehicle. In contrast, the third vehicle receives the acceleration forecasts from its TV that is operating in CACC mode, thereby reacting to the braking maneuver faster. The peak deceleration stays at the same level as for vehicle two, thus vehicle three does not amplify perturbations further. The peak deceleration values in the third and fourth vehicles are also reduced compared with the previous vehicle. This observation underscores the benefits to CACC performance by transmitting acceleration forecast data elements.

Another important aspect during a deceleration maneuver was the overshoot in the vehicle speeds in the following vehicles. From Figure 8*a*, it can be observed that there are significant fluctuations in both these parameters in the case of ACC. For instance, the minimum speed for the fifth vehicle approached 0 m/s. On the other hand, the CACC system showed a much more stable response with speeds not dropping below 9 m/s and a much smaller range of speeds overall.

Optimization of Time-Gaps

The simulation scenario considered for this analysis involves a CACC string comprising three compact cars followed by two large SUVs, with a time-gap of 0.5 s. A perturbation is introduced to the string in the form of a moderate deceleration maneuver of magnitude 3.0 m/s^2 and duration of 5 s by the lead vehicle. Figure 9 is a truncated plot of the temporal variations in the relative time-gaps of vehicles for the OVLA and MVLA with behavior estimation, respectively. Each curve corresponds to the time-gap variation for a pair of contiguous vehicles in the string. It can be clearly seen that the variations in the time-gap for the latter approach are more constrained, thereby implying more stability in the time-gaps for the MVLA approach.

A Proposal for CACC Messaging

The V2V messages to enable CACC functionality were based on the latest SAE J2735 (*15*) data dictionary and the BSM. Additional data elements for enhanced CACC performance were identified during the course of the presented research work, as shown in Table 1. The key point is the transmission of the information about the predicted

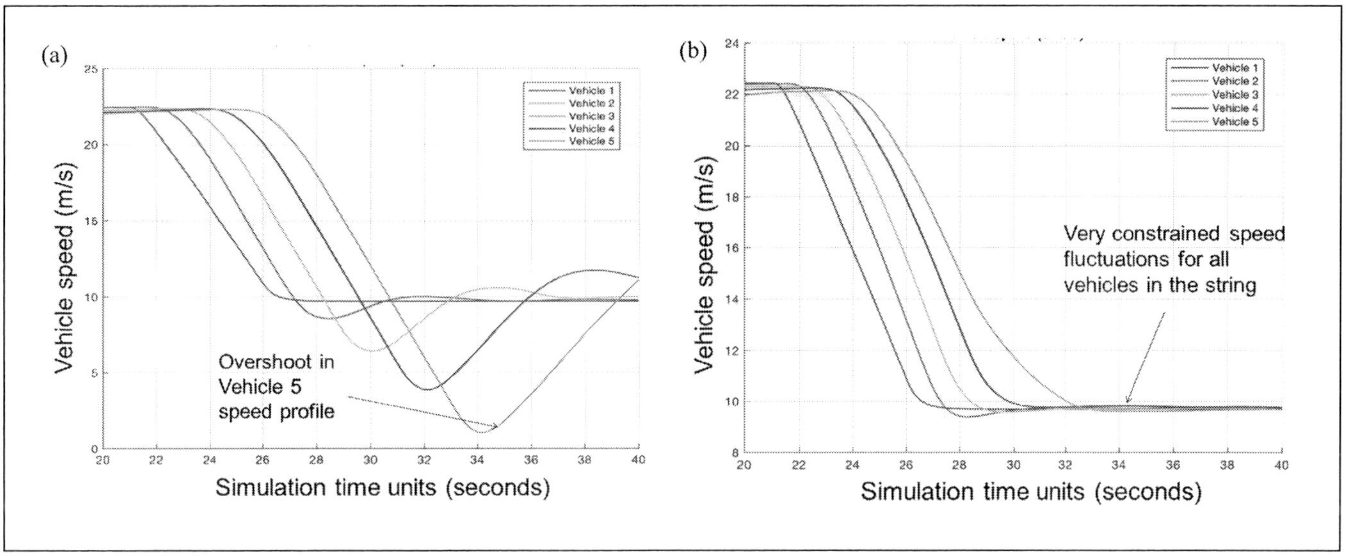

Figure 8. Vehicle speed profiles for (a) ACC in and (b) CACC in.

future motion of an HV to other vehicles. Because a longitudinal controller commonly controls vehicles through an acceleration interface, this is implemented as an "acceleration forecast" of the HV. Acceleration commands are being transmitted over the air as soon as the longitudinal controller issues them to the brake and engine controls in the HV, effectively making them acceleration forecasts because of the vehicle response time in implementing the command. To model differences in vehicle response performance, a first-order lag model was successfully used. The vehicle-dependent time constants of that model are also transmitted, thereby enabling following vehicles to estimate the acceleration response of the HV. This results in an enhanced performance for the CACC system. This approach was validated in the previous section, especially by Figure 7, underscoring the importance of transmitting these additional data elements. A high-level proposal for a novel message to enable CACC behaviors is described below.

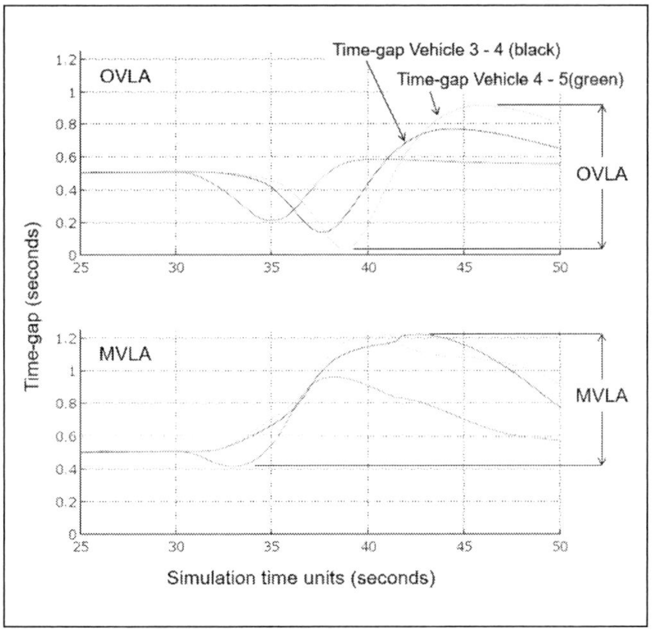

Figure 9. Temporal variations in the time-gaps of the vehicles in the CACC string for the one-vehicle look-ahead (OVLA) and multi-vehicle look-ahead (MVLA) approaches under moderately harsh deceleration maneuver.

Conclusions

The concept of augmenting an ACC framework with vehicular communication capability and additional processing modules to realize CACC functionalities was the crux of the paper. A novel approach of integrating relevant road test data to a simulator to establish the benefits of CACC in vehicle strings was presented. In addition to quantifying the anticipated performance enhancements of CACC for a vehicle string (in the form of reduced system latencies, situational awareness, etc.), the paper also developed modules that will aid in realizing CACC behavior. Additional data elements to further smoothen the longitudinal maneuvers were identified. Future work includes an analysis of the messaging requirements for CACC, integrating the CACC modules into the test vehicles, and conducting road-based experiments to validate the algorithms. Lastly, it is remarked that the CACC modules can be adapted to other sensor-based, longitudinal control systems such as the ones envisioned for highly automated vehicles.

Table 1. Additional Data Elements Required for CACC

Data element	Purpose	SAE J2735 datatype
CACC state	• To enhance awareness of the capabilities of vehicles in the string / neighboring vehicles • Seven states: Manual, Cruise Control, ACC and CACC, higher levels of automation (sensor-based), higher levels of automation (V2V data fusion-based), or Manual Override	Longitudinal control state
Acceleration forecast	• "Intended value" when the HV issues a command for brake, throttle, and / or engine control	Acceleration
Tau	• Time constant of first-order lag models for the engine control acceleration / deceleration schedules[a]	Time constant
Target vehicle temporary ID	• Aid with the in-lane assessment of vehicles • When the HV is following another DSRC-equipped vehicle, and has validated that this vehicle is in the same lane (through its object fusion algorithm), it will transmit the identifier of that vehicle as its TV's temporary ID • Essential for the linked-list algorithm that allows vehicles to compile a list of vehicles that are part of the same string	Temporary ID

[a]The times for acceleration and deceleration were between 400–900 ms and 200–1200 ms, respectively.

Acknowledgments

The project was conducted by the Crash Avoidance Metrics Partners LLC (CAMP) Vehicle-to-Infrastructure (V2I) Consortium. The participating companies in the V2I Consortium are Ford, General Motors, Hyundai-Kia, Honda, Mazda, Subaru, Volvo Technology of America, and Volkswagen Group of America. This material is based upon work supported by the U.S. Department of Transportation under Cooperative Agreement No. DTFH6114H0002.

Author Contributions

The authors confirm contribution to the paper as follows: study conception and design: J.-N. Meier, A. Kailas, O. Abuchaar, M. Abubakr, R. Adla, M. Ali, G. Bitar, R. Deering, U. Ibrahim, P. Kelkar, V. Vijaya Kumar, E. Moradi-Pari, J. Parikh, S. Rajab, M. Sakakida, and M. Yamamoto, J.-N. Meier, A. Kailas, O. Abuchaar, M. Abubakr, R. Adla, M. Ali, G. Bitar, U. Ibrahim, P. Kelkar, V. Vijaya Kumar, E. Moradi-Pari, J. Parikh, S. Rajab, M. Sakakida, and M. Yamamoto; simulations and data collection: J.-N. Meier, A. Kailas, O. Abuchaar, M. Abubakr, R. Adla, M. Ali, G. Bitar, R. Deering, U. Ibrahim, P. Kelkar, V. Vijaya Kumar, E. Moradi-Pari, J. Parikh, S. Rajab, M. Sakakida; and M. Yamamoto; analysis and interpretation of results: A. Kailas, J.-N. Meier, O. Abuchaar, M. Abubakr, R. Adla, M. Ali, G. Bitar, R. Deering, V. Vijaya Kumar, E. Moradi-Pari, S. Rajab, and M. Sakakida; draft manuscript preparation: Y. Aravind Kailas, Pascal Amar. All authors reviewed the results and approved the final version of the manuscript.

References

1. Mahmassani, H., H. Rakha, E. Hubbard, and D. Lukasik. *Concept Development and Needs Identification for Intelligent Network Flow Optimization (INFLO)-Assessment of Relevant Prior and Ongoing Research.* Publication FHWA-JPO-13-011. Federal Highway Administration, US Department of Transportation, Washington, D.C., 2012.
2. Arnaout, G. M., and S. Bowling. A Progressive Deployment Strategy for Cooperative Adaptive Cruise Control to Improve Traffic Dynamics. *International Journal of Automation and Computing.* Vol. 11, No. 1, 2014, pp. 10–18.
3. Michaelian, M., and F. Browand. *Field Experiments Demonstrate Fuel Savings for Close-Following.* Publication UCB-ITS-PRR-2000-14. California PATH Program, UC Berkeley, Berkeley, CA, 2000.
4. Amoozadeh, M., H. Deng, C.-N. Chuah, M. H. Zhang, and D. Ghosal. Platoon Management with Cooperative Adaptive Cruise Control Enabled by VANET. *Elsevier Vehicular Communications*, No. 5, 2015, pp. 110–123.
5. Parikh, J., O. Abuchaar, E. Haidar, A. Kailas, H. Krishnan, H. Nakajima, M. Maile, J.-N. Meier, S. Rajab, Y. Sharrab, S. Siko, J. Thompson, M. Yamamoto, and R. Deering. *Vehicle-to-Infrastructure Program Cooperative Adaptive Cruise Control.* Publication FHWA-JPO-16-257. Federal Highway Administration, US Department of Transportation, Washington, D.C., 2015.
6. Rajamani, R., H.-S. Tan, B. K. Law, and W.-B. Zhang. Demonstration of Integrated Longitudinal and Lateral Control for the Operation of Automated Vehicles in Platoons. *IEEE Transactions on Control Systems Technology*, Vol. 8, No. 4, 2000, pp. 695–708.
7. Rajamani, R., E. Chan, and E. Coelingh. An Experimental Comparative Study of Autonomous and Co-operative Vehicle-Follower Control Systems. *Transportation Research Part C: Emerging Technologies*, Busan, South Korea, 2001, pp. 1–11.
8. Robinson, T., E. Chan, and E. Coelingh. Operating Platoons on Public Motorways: An Introduction to the SARTRE Platooning Programme. *Proc., 17th World Congress on Intelligent Transport Systems*, Busan, South Korea, 2010, pp. 1–11.
9. Fritz, H., A. Gern, H. Schiemenz, and C. Bonnet. Chauffeur Assistant: A Driver Assistance System for Commercial Vehicles Based on Fusion of Advanced ACC and Lane Keeping. *IEEE Intelligent Vehicles Symposium*, Parma, Italy, 2004, pp. 495–500.

10. Ploeg, J., B. T. M. Scheepers, E. van Nunen, N. van de Wouw, and H. Nijmeijer. Design and Experimental Evaluation of Cooperative Adaptive Cruise Control. *Intelligent Transportation Systems Conference*, 2011, pp. 260–265.
11. Güvenç, L., I. M. C. Uygan, K. Kahraman, R. Karaahmetoglu, I. Altay, M. Senturk, M. T. Emirler, A. E. H. Karci, B. A. Guvenc, E. Altug, and M. C. Turan. Cooperative Adaptive Cruise Control Implementation of Team Mekar at the Grand Cooperative Driving Challenge. *IEEE Transactions on Intelligent Transportation Systems*, Vol. 13, No. 3, Sept. 2012, pp. 1062–1074.
12. Lu, X.-Y, S. E. Shladover, A. Kailas, and O. Altan. Proposed Message Set for Cooperative Adaptive Cruise Control using V2V in Real Traffic. *Vehicle-to-Vehicle and Vehicle-to-Infrastructure Communications: A Technical Approach* (Hu, F., ed.), Taylor & Francis LLC, CRC Press, Boca Raton, FL, May 2017.
13. SAE J2945/1. *Surface Vehicle Standard: On-board Systems Requirements for V2V Safety Communications*, 2017.
14. Ploeg, J. *Analysis and Design for Controllers for Cooperative and Automated Driving*. Eindhoven University of Technology, Eindhoven, the Netherlands, 2014.
15. SAE J2735. *Dedicated Short Range Communication (DSRC) Message Set Dictionary*, 2016.

The Standing Committee on Vehicle-Highway Automation (AHB30) peer-reviewed this paper (18-03224).

Any opinions, findings, and conclusions or recommendations expressed in this publication are those of the authors and do not necessarily reflect the view of the U.S. Department of Transportation.

Sage Publications USA
sagepub.com

ISBN 978-1-5108-7827-3

CURRAN ASSOCIATES INC.
proceedings.com

International Conference on Ice Class Vessels 2020

Online
4 November 2020

The Royal Institution of
Naval Architects